WASTED

WASTED

How We Squander Time, Money, and Natural
Resources—and What We Can Do About It

Byron Reese &
Scott Hoffman

CURRENCY
NEW YORK

Published in the United States by Currency, an imprint of Random House, a division of Penguin Random House LLC, New York.

CURRENCY and its colophon are trademarks of Penguin Random House LLC.

Hardback ISBN 978-0-593-13518-1
Ebook ISBN 978-0-593-13519-8

Printed in Canada on acid-free paper

crownpublishing.com

9 8 7 6 5 4 3 2 1

First Edition

Book design by Caroline Cunningham

Contents

PART 3: THE SCIENCE OF WASTE

PART 4: THE PHILOSOPHY OF WASTE

Introduction

Imagine you have ordered something from an online retailer—a computer cable, perhaps. The delivery driver drops the box off at your house, and you waste a minute or two finding a pair of scissors to cut through the multiple layers of tape.

You pull the item out of the box, which will later be thrown away and wasted, and open it up, discarding that packaging as well. Finally you uncover the computer cable you ordered.

But alas, after wasting several minutes trying to get it to work, you realize it's the wrong one. You waste more time figuring out which cable you do, in fact, need. You then waste time repackaging the cable and initiating a return online, which takes a few more minutes.

A delivery driver picks the box up from your home, and it is flown back to some return center, where yet another person opens it, credits your account, and either sends the product back to the manufacturer or throws it in a bin of returned merchandise that might be purchased by another company, which will then have to sort through it all, classify it, and try to sell it. Then . . . well, you get the idea.

And then imagine the replacement cable you ordered is defective.

It is wearisome to think through the ramifications of even a single misordered cable, and yet snafus like that happen to most of us on a regular basis. These mistakes, in aggregate, generate an unfathomable amount of waste.

But the cable example actually represents a trivially small amount of waste compared to what awaits you elsewhere in this book. If you zoom the lens out, you see a larger expanse of waste, such as the waste of the time you spend in traffic, the wasted heat your car generates, even the waste inherent in warfare. All of these examples represent the types of waste we'll explore further. But even in examining these sources of waste, we hardly scratch the surface. If we pull back even further, we see a larger world of waste: the wasted potential of a life lived in poverty, the waste of a person who dies needlessly, and more.

Waste is everywhere. And once you train your eyes to look for it, you will see it all around you. The next time you fill up your car with gasoline and inevitably drip a few drops on the ground, you will recall that, collectively, these drops add up in volume to an oil spill the magnitude of the one from the *Exxon Valdez*. Every single year. In the United States alone. Then, if you add up the amount of gasoline spilled when people fill up their lawnmowers and garden equipment, you get another *Exxon Valdez* oil spill every single year.

When a tanker ruptures and spills its load or a pipeline bursts, it's headline news. But a fraction of an ounce of spilled gasoline barely registers. Sure, we recognize a small loss, but we never really think about how all the waste and inefficiency in our lives add up, and what it all *means*, ultimately.

In writing this book, we went on a journey to try to understand the waste inherent in modern life. We both soon realized how little we knew about so many things related to waste. So we started posing all kinds of questions:

- Is it better to wipe up a spill in the kitchen with a paper towel or a damp rag?
- How would the world be different if there had never been any war?
- Does the money you save buying bulk goods offset the amounts of those goods that spoil?
- What portion of the twenty-four hours in a day do we waste?
- When you go on a diet and lose weight, where does that mass actually go?

Some answers are easy to figure out. With regard to the paper towel question, a damp rag wastes less, unless you wait until the tap water gets warm or you wash the rag after every use. If either condition is true, use the paper towel.

And regarding weight loss, 84 percent of any weight you lose gets expelled in the form of carbon dioxide when you exhale and 16 percent is expelled as water through urine, sweat, and tears. The question of whether carbon dioxide from humans is waste comes up later in this book. Urine is (assuming you're not the type to recycle the minerals in it), but sweat is less clearly so, since it serves the obvious function of regulating body temperature. We'll let you decide the extent to which tears are waste.

Incidentally, none of the weight is "burned," or directly turned into power or heat for your body. Your body gets its warmth and energy from breaking down bonds between molecules. As a result, *every* bite you eat eventually comes back out as air and water. (The solid waste you excrete comes from food you ate recently, and so was never "weight" to begin with.)

Other areas of waste, however, are more nuanced and counterintuitive. When you examine things that seem to be obviously wasteful, digging a little deeper reveals that there is often a method to what appears to be human madness—for instance, why it makes sense to ship rocks halfway around the world to extract aluminum from them, or to print hats and T-shirts announcing Super Bowl winners who never won the big game.

No one *wants* to waste, and people spend a great deal of mental energy trying to avoid it, which is laudable. But the world is full of well-meaning attempts to avoid waste that actually cause more waste than they prevent.

A recent documentary highlighted a family of "freegans" who decided to live for a year off food that had been thrown away. Their thinking was that if they consumed food that might otherwise go to a landfill, they would be saving the world from waste and inefficiency. But the gasoline that powered the car they used to drive from place to place to *get* those "waste" calories is a nonrenewable resource that they used to retrieve a renewable one, food. Then there's the time they spent driving around, the depreciation of the car itself, and so forth.

One of the major themes of this book deals with what is seen and what is unseen. When we think of waste, it's easy to picture all the food that's thrown away. It's much harder to see the second- and third-order effects of trying to avoid waste in the first place. In many cases, understanding waste requires diving several layers deeper to understand that a behavior that seems wasteful (or efficient) at first glance is anything but.

Simple questions, like paper versus plastic or flying versus driving, are devilishly complex. Recycling paper, it turns out, is great—if a recycling truck comes around to get it. Drive those old newspapers across town and you have done more harm than good.

Making these determinations often involves many trade-offs, such as whether methane or carbon dioxide is worse for the environment, or whether it is better to consume a large amount of freshwater or a small amount of fuel.

It really does depend on what the fuel is and where the water comes from, and there is a different answer in Iceland than in Islamabad. And to top it off, we can't rely on our intuition to guide us because so much is counterintuitive. Who would ever guess, for instance, that the family dog has a bigger carbon footprint than the family SUV?

How should this knowledge impact our behavior? And who has time for all this? Who wants to learn a thousand rules governing all the minutiae of their life, right down to whether they should get the plastic or glass bottle of ketchup—only to discover that most rules are situational, not universal?

We often face decisions like "Should I choose the path that adds most to climate change, the one that supports an oppressive regime, or the one that contributes to the extinction of species in the Brazilian rain forest?" Unlike the damp rag example, where an answer is pretty clear, we are left unsure of every choice, and end up feeling vaguely guilty about whichever one we make.

An additional layer of unease and guilt complicates the picture further when we consider the vast amount of resources consumed by some and the extreme dearth available to others. This aggregate feeling of confusion and paralysis is real and palpable.

For many, the solution to the confusion is to try not to think about it. We fall back on "Well, one person can't really make a difference anyway." But just like those few drops of gasoline that add up to an *Exxon Valdez,* our collective actions do matter, and it turns out that most of us want to reduce waste in all forms. And if we simply knew the "best" choice, most of us would choose it.

This is the paradox we find ourselves in. We live in the Information Age, where, in theory, most everything is knowable. And yet, in practice, it isn't really. The level of interconnectedness in modern life can be a barrier to understanding the interplay between so many complex systems. Thomas Hobbes suggested that "hell is truth seen too late," and yet our challenge is seeing truth at all.

Some argue that, in the end, we should all just try to reduce consumption. Try to get by on less. And in this advice there is unquestionably much truth and wisdom. But it isn't much of a path forward. How much less? And less of what? Fewer antilock brakes or prenatal exams? Less refrigeration or cancer-curing drugs? Less pizza? Or maybe just fewer toppings? Most Westerners can't, or won't, live the lives of ascetics, eschewing all consumption and worldly pleasure, so that's clearly not the answer, either.

These issues are baffling, but it is important not to turn cynical, nor throw your hands up in frustration at the utter complexity of it. There *are* ways to live that minimize, or at least reduce, waste. But how much waste is it even possible to eliminate?

This question transformed the topic of this book. The book began as a study of waste. But the questions we found more engaging were related to what a world *without* waste would look like. No waste whatsoever. No wasted paper, no wasted food, no wasted time, no wasted minds, no wasted lives.

What's the purpose of such a question? Certainly, it is theoretical. We can't live in a world without waste unless we have perfect knowledge of almost everything, and we never will. But one of the things this book is quite interested in is understanding the upper limit of what *is* possible and how we might move toward it.

Imagine a horizontal line. On the far right is a world of perfect efficiency, where there is no waste—everything gets used to its fullest potential. On the left, a world of pure waste—all time is wasted, all money is wasted, everything. As we researched this book, we came to understand we are far, far to the left, very close to a world of pure waste. Why? How can it be?

We have come to believe that we waste so much because we

don't understand waste very well. How do inefficiency and misinformation come together to give us the wasteful world we find ourselves in?

The message of this book is that we can move along this continuum toward a world without waste, where we are all happier and healthier.

This is not a policy book. While we'll weigh in on policy issues, like whether deposit schemes on bottles work and the effects of outlawing plastic bags, this book is more about science, economics, and systems than political choices. It doesn't try to guilt or shame anyone into anything; rather, it is designed to take you on the same journey that we went on, of understanding why waste happens and imagining a world without it.

What Is Waste?

Socrates once said—or at least is alleged by Plato to have said—"The beginning of wisdom is the definition of terms." While definitions seem like a straightforward place to start, for many complex concepts, they are often elusive.

For instance, there is no universal definition of life, which seems odd since it is the biological fact with which we have the most firsthand experience. Likewise, there are several definitions of death. The definitions of time and space are hotly debated, and family squabbles over definitions of intelligence have ruined many a Thanksgiving dinner.

In fact, most of life's most complex concepts are more intuited than defined. Right and wrong, love and hate, family and friends, home, health, faith, and art are nebulously defined in most of our minds, and your notion of them is probably different from that of the people around you.

That brings us to waste. What is it? At first glance, it might not seem such a difficult concept to define. We all know more or less what we're talking about, right? Maybe we should skip definitions altogether. But Socrates would shake his head disappointingly at this reasoning and find it a sloppy way to begin, and rightly so.

Waste is actually a challenging concept to wrap one's mind around, primarily because we use it in so many contexts. Waste can be an unwanted byproduct, or something otherwise useful that gets damaged or destroyed. Our bodies can waste away, and we might speak of "wastelands," both literally as geographic features and metaphorically, as former FCC commissioner Newton Minow once described the television programming landscape. But you can also waste time and potential.

Sometimes, however, we use the word "waste" to describe things that aren't really waste at all. For instance, when it comes to hydrogen fuel cells that produce power, the only byproduct is water. Although this water is colloquially referred to as "waste," few would regard it as so—certainly not in the way that we look at the carbon dioxide that results from burning fossil fuels to produce energy.

Let's start by considering an expansive definition of waste. Can waste be thought of as not achieving the most optimal outcome in any situation? In any given twenty-four hours, could we count up the minutes spent doing unnecessary or unwanted tasks and call that time "wasted"? Could we count as waste every extra step we had to take to get where we wanted to go or every minute we spent sitting in traffic?

We could, but this way of looking at waste would be inadequate for two reasons. First, it is only an individual understanding—a relative rather than absolute measure. You may have been stuck in traffic for a minute because an ambulance was speeding a heart-attack victim toward lifesaving care. From a societal view, the allocation of everyone's time in traffic may or may not have been optimal; we just don't know. Second, in some cases certain events are unavoidable, given the limits of human power and understanding, and so cannot be characterized as waste. You may have had to take those extra steps because a sinkhole appeared in the middle of the sidewalk, forcing you to walk around it.

While the broad definition of waste we've outlined has some problems, it is useful, for it highlights three elements that need to be incorporated in our understanding of waste. First, waste is undesirable. In other words, the pleasant aroma that wafts from your oven while baking chocolate chip cookies isn't waste, while the sulfurous reek from an industrial smokestack is. Second, it must incur a cost of some kind *without some net offsetting benefit*. Thus, we can't say that the night you had trouble sleeping because the music at your neighbor's rave kept you up is *necessarily* waste, because there *was* a benefit—just not to you. And third, waste must be avoidable. If it isn't, such as the damage to an island town caused by a tsunami, we can call the event a tragedy, but it isn't waste.

With these pillars—that waste is undesirable, incurs a net negative cost, and can be avoided—we have another question: What sort of thing is waste? The word itself can, after all, be a noun, a verb, and an adjective. Further complicating matters, waste can involve the creation of something or the destruction of something (or both). In this book, you will mostly find the word "waste" used in its noun form, and we use it to mean both a wasted item itself as well as the unrealized difference between what occurs and what could potentially occur.

At certain points in the book, we examine outcomes that are presently unavoidable, but for which there are theoretically possible technological solutions. These will be self-explanatory.

This understanding of waste, then, is sufficiently broad to include smallpox, oil spills, and premature death. We must be careful, however, not to play the waste card too quickly. Can you really waste water, since it is never really destroyed or created? Does wasting paper matter if it's a renewable resource? Is the slacker who spends his days smoking pot really wasting his life if that is what he wants to do? If you *could* get by on less sleep, but don't, are you wasting your time when you get those extra hours? These are all questions that will be examined over the course of the book.

But there is one final aspect of waste that bears examination. From a certain perspective, if you zoom the camera out far enough and view the universe on a twenty-billion-year timeline, then one could argue that everything is fundamentally pointless and thus *all* activity is wasted when viewed against that cosmic background. Everything we do is, as Shakespeare's Macbeth described it, a tale "told by an idiot, full of sound and fury, signifying nothing." When viewed in this manner, even our very best actions in life are little more than polishing brass on a sinking ship.

But this view of waste is only true if there is no inherent and lasting value in human existence *in and of itself*. And that nihilism warrants some consideration.

In defining waste, one of our key attributes is that it is undesirable. Implicit in that idea is that it is undesirable *to humans*. We take antibiotics to slaughter billions of bacteria with little regard for their wasted lives, cut short by our ruthless modern medicine. For that matter, many of us eat bacon, which presumably a pig would regard as a waste of its life. But we seldom count either as waste.

Most of us live our lives as though only humans can impute value to things. A beautiful sunset is not itself beautiful; it only becomes so because we declare it to be beautiful. Even the lives of animals have worth only because we assign it to them, somewhat capriciously, based on whether we perceive them as cute or useful to us. Thus, we bug-bomb roaches en masse but pass laws protecting baby seals. But we do regard human life, all human life, as inherently worthwhile, which is the basis for the idea of human rights. That's the perspective this book adopts when discussing waste.

So why dip into philosophy at all? Because, as Socrates suggests, we must include in our definition of waste that it is *purely a value judgment*. A picturesque landscape blighted by litter is only bad to the extent we regard it as so. Thus, all waste must be seen through a hierarchy of value, arranged like a tower. Many things sit atop this edifice. A few that come to mind include individual liberty, Mother Earth, an end to want and misery, the amicable companionship of all living creatures, a person's children or their country, or the will of their God.

Because we all have different value systems, we may not always agree if something is waste or not. This system of relative values may seem like it makes this book impossible, but the situation is not so bleak. Jack Kennedy put it best when he said, "Our most basic common link is that we all inhabit this small planet. We all breathe the same air. We all cherish our children's future. And we are all mortal." For most of us, our value systems overlap enough that we can live in peace and harmony with each other. If this were not the case, there would be no human civilization.

Thus, in writing this book, we assume a shared set of core values: that we all prefer clean air, beaches without needles washing up on them, a minimum of human suffering, and so on. We also assume that human existence is inherently of value—that even after the cold death of the universe, the value of a human life survives.

In short, it matters that each of us is here.

PART 1

Waste and Our Planet

In a Million Years, They Will Know Us by Our Polystyrene Foam

In a million years, humanity as we know it will be long gone. No matter your philosophy, cosmology, or theology, we won't be here. We might have evolved into *Homo somethingelsius*, uploaded ourselves to computers, or augmented ourselves into androids. We might have succumbed to an ecological catastrophe, blown ourselves into oblivion, or been hit by an asteroid; we might have been wiped out by Wolf-Rayet star WR 104, which will have gone supernova by then, or abandoned this planet and gone to the stars ourselves. Or we might have been overthrown by a band of intelligent apes with a charismatic leader named Caesar, or any of a hundred other outcomes. But rest assured, we won't be here in any way that resembles our lives today.

However, the polystyrene foam cup you drank hot coffee from this morning *will* still be here, buried somewhere, probably still quite recognizable, even usable.

Saturn's rings will have vanished. Polaris will no longer be the North Star, and the night sky will look entirely different, missing the Big Dipper, Orion, and the rest of the gang. But that white foam cup will still be around. For an item whose useful life is measured in minutes, that's pretty impressive.

But the story of that cup, its origin and destiny, is a bit more complicated than that simple narrative. So let's take Lewis Carroll's advice and begin at the beginning, with an organic compound called styrene, which has the chemical formula $C_6H_5CH=CH_2$. It's a liquid that is manufactured from petroleum in great quantity—roughly thirty million tons of the stuff a year. It can be used to make latex or synthetic rubber, but most often it's polymerized into polystyrene, a plastic.

Polystyrene, in turn, can be made into all kinds of hard plastic items, like the housing for your smoke alarm or your silverware tray. It can also be heated to boiling, which makes it expand to forty times its original size, turning it into expanded polystyrene foam, or EPS. Expanded forty-fold, EPS is about 97 percent air, making it an excellent insulator and cushioning material, and therefore ideal for certain applications, such as disposable coolers, food takeout containers, packing peanuts, and the aforementioned coffee cup.

In this form, EPS is colloquially referred to as Styrofoam, but that's not technically accurate. Styrofoam is a trade name for a product made by DuPont, which is made of XPS—that is, extruded, not expanded—polystyrene. Real McCoy Styrofoam is used in construction, not in single-use applications. The folks at DuPont really don't like people using their product name in a generic fashion to apply to a different kind of product. They go so far as to maintain a page on their website called "Styrofoam is not a cup" that explains that it is EPS that is used to make cups and containers—a technology completely different from real-deal Styrofoam.

EPS has been cast as the villain of our throwaway culture for quite some time. McDonald's, for instance, announced its intention to stop using foam packaging way back in 1990. Since then, while EPS is never explicitly depicted as laughing maniacally while twirling its mustache, its villainy is assumed.

Its list of character flaws is damning. It doesn't biodegrade, but it does degrade when exposed to light, entering the food chain and killing animals. It can't be easily recycled. Worse, recyclers hate it because when EPS gets broken up in the recycling stream, it contaminates other items, making it impossible to recycle them at all. Even if EPS could be economically recycled, it is frequently covered in food, complicating matters. It weighs so little that it gets blown into the ocean, it clogs up drainpipes, and both its manufacture and its destruction are often bad for the environment.

Because EPS is mostly air, and because the world uses so much of it—around 15 million tons annually—the volume of EPS we produce is immense. Some estimates peg the volume of landfills taken up by EPS at as much as 30 percent, although other estimates suggest a smaller number. EPS's volume also gives us less incentive to recycle it. A semi-trailer loaded with aluminum—well, you can put your kid through college with that. But a truckload of EPS is pretty much a truckload of air with a little plastic sprinkled in. It's not worth the fuel it takes to get it to a recycling plant. So hated is EPS that it has been declared a material non grata and banned in many countries around the world, as well as a plethora of cities and states in the United States.

And yet, like many charming movie antiheroes, EPS does have virtues, at least compared to alternative forms of packaging. It requires fewer resources to make, and its high strength-to-weight ratio makes it less expensive to ship compared to other alternatives, which saves on energy use and CO_2 emissions. For the purposes it is usually used for—insulating food, packing electronics, and so on—there are few materials that work as well, and none at the very low price of EPS. These virtues are meaningful, and all true. But the criticisms of EPS are also true, or at least partly true.

Let's dig deeper into a few. First, does EPS degrade? In a land-fill, not really. But of course, very little does. Landfills keep out light, which is EPS's great nemesis. But when floating in the ocean, yes, it does break down. And when the polymer EPS breaks back down into the monomer styrene, it can do damage to the ecosystem. It crumbles, so the pieces can end up getting in-gested by aquatic animals.

Can EPS be recycled? As mentioned earlier, yes, but . . . Its denser cousin, just plain ol' polystyrene (not expanded), is recy-clable. You might see the recycling code 6 stamped on the under-side of polystyrene items, like plastic cups. In the United States, more than a hundred businesses do recycle EPS—but most don't want to deal with it. The logistical issues of transport and con-tamination can be solved, but doing so generally costs more than the recycled material is worth. And since it's a fossil fuel byprod-uct, the value of secondary production (recycling) is as volatile as the price of oil.

Thus, whether EPS will ever be recycled at scale is still an open issue. New York famously banned EPS in 2013. The city was promptly sued, and in 2015 the ban was temporarily put on hold while the economic viability question was studied. In 2017 the city released a report called "Determination on the Recyclability of Food-Service Foam" and found that "Food-Service Foam compacts in collection trucks, breaks into bits, and becomes cov-ered in food residue, making it worthless when it arrives at the material recovery facility ('MRF'). It then blows throughout the MRF, is missed by manual sorters, mistakenly moves with the paper material and contaminates other valuable recycling streams, namely paper, which can be the most consistently valu-able commodity in a recycling program." The report's conclusion was that EPS could not be recycled economically. A judge in the case found the New York study to be compelling, and ultimately allowed the ban to become law.

Are bans the right answer? EPS proponents argue that bans are fundamentally counterproductive and driven by politics. It's easier for a politician to look like an environmental hero, they argue, than to do the hard work of getting curbside recycling of EPS into place.

But the elephant in the room is the question "If we ban EPS, what will people use instead?" It is quite possible that the substitutes—wax-coated cardboard or rigid polystyrene, for instance—might cause more environmental harm than the EPS they are offsetting. If people end up just swapping one material for another, then there is no net reduction in the amount of trash. And the new materials, contaminated with food, may be no more recyclable than the EPS. This is the reason that many environmentalists reject one-off bans, such as on EPS, plastic bags, and straws, in favor of more sweeping legislation relating to all plastics. The big issue, they argue, is that we have all these single-use plastic items that end up in landfills, and that is the proverbial ball we should keep our eye on. And it isn't even so much that they end up in landfills that is the problem; rather, it's that single-use items are made to begin with. Tackling any particular plastic item is at best a PR ploy and at worst a distraction from the real problem.

There are only three ways out of this conundrum. One is a widespread change in behavior away from single-use items of all kinds, plastic and otherwise, perhaps by stigmatizing them the way cigarette smoking was stigmatized. The second is to recycle them regardless of the cost, accepting that perhaps doing so cannot ever be profitable; a truckload of EPS may simply have negative economic value. Finally, technology might come to the rescue. A number of start-ups are exploring ways to overcome the hurdles to recycling EPS described in this chapter, or to find new materials that do the job of EPS in a more sustainable manner. Researchers at Washington State University, for instance, claim to have made a better insulator than EPS out of nanocrystals of cellulose, which they say is the most plentiful plant material available. There are some interesting potential biological solutions to the problem as well. In one recent study, it was shown that mealworms fed on diets of pure EPS thrived just as well as those fed a diet of bran, although that can also be interpreted as a pretty scathing indictment of bran. Regardless, it turns out that mealworms' gut microbes break down EPS entirely.

Would a world without waste have EPS? Probably. There are many applications of EPS that aren't problematic, such as building insulation. But for the most part, the use of single-use EPS containers is highly wasteful, and less wasteful options exist.

Are Landfills the Green Choice?

Every year, the more than seven billion people on earth produce about 2 billion tons of garbage—municipal waste, if you prefer the technical term. That works out, roughly speaking, to 1.5 pounds per person a day. This number varies widely by country in a fashion that is correlated to per capita GDP. Put simply, the more money you have, the more trash you create. In the United States, for instance, we produce about triple the average, or just under 5 pounds per person per day.

About one-fifth of all the world's waste is recycled or composted. Of the remaining 80 percent, the World Bank estimates that half ends up dumped openly or burned, while the other half goes to landfills, which are often viewed as the villains of the trash world. But is this a fair characterization?

After all, as much as we hear about how landfills are all filling up, few people come across them in their day-to-day lives. They aren't around every corner, overflowing with garbage, blighting landscapes. In fact, it turns out there are only about two thousand active landfills in the United States, a number that falls over time as each new landfill becomes ever bigger. But just because there aren't many of them and they're largely kept out of sight doesn't make them benign. No, the story of landfills is a bit more complicated than that.

First off, how do landfills work? The idea behind them is pretty simple: You dig a big hole, and line it in such a way that nothing can leak out of it. Then, every day, you compact trash and dump it in the hole. Covers are placed over it to keep odors in and to keep garbage from blowing away. This goes on until the hole is filled.

But there's a complication to this straightforward story. Landfills create two waste products. The first is called leachate, the water that collects at the bottom of the landfill. It consists of rain that has fallen onto the landfill, then filtered down through all the junk. It's some really nasty stuff. Think about it. That water goes through old batteries, dirty diapers, discarded medicines, motor oil, rotting food, obsolete electronics, and all the other stuff that finds its way into our trash. It then settles on the bottom, where it festers. Get too much leachate in your landfill, and the lining can bust. So the leachate has to be pumped out and treated, which isn't easy but is doable. This process requires energy and incurs cost.

However, that cost can be offset by the second waste product, landfill gas. Landfill gas is a combination of methane and carbon dioxide, which is released as items in the landfill decompose. Landfills are designed to *prevent* their contents from biodegrading for this very reason. However, all kinds of highly biodegradable items like food find their way into the landfill, where they're broken down by bacteria and decompose.

As a greenhouse gas, the carbon dioxide produced by landfills is problematic, but the methane is arguably worse, particularly in the short term. In the atmosphere, methane can trap roughly thirty times more heat than carbon dioxide, and landfills are the third-largest human-made source of its emission in the United

States. (Source number one is petroleum and natural gas production, and source number two is what the U.S. Environmental Protection Agency, or EPA, calls "enteric fermentation," or cow burps, which, while technically not human-made, is a human-caused problem. Without people, there would be far, far fewer cows.)

Since 1990, the United States has required that landfills deal with the methane in some fashion. At the most rudimentary level, some landfills simply flare off the methane—they light it on fire. Burning 1 pound of methane creates 2.75 pounds of CO_2 (to burn the methane, we have to add environmental oxygen, which is why the products of combustion weigh more than what you're burning in the first place).

However, there is a much better thing to do with landfill gas. If your landfill is big enough, you can use it to generate power. Lots of it. The methane in landfill gas contains about as much energy as low-grade coal but burns much more cleanly. Burning the methane for energy not only displaces the dirtier fuels that were previously used to evaporate the leachate but also can provide a surplus of usable energy.

The largest landfill in the United States, Puente Hills, located near Los Angeles, produces 500 cubic feet of landfill gas *per second,* which, when burned, provides enough surplus electricity to power forty thousand homes. The EPA suggests that across the United States, by displacing dirtier sources of electricity and reducing methane emissions, burning landfill gas provides environmental benefits equivalent to a forest the size of California.

By turning landfill gas into energy, we take one of the clearest-cut examples of how to create value from what seems like pure waste. While landfills are no panacea, until we can reduce the volume and nature of waste at the source, there are certainly worse things to do than to put municipal waste in well-constructed landfills.

Of course, there are still major drawbacks to landfilling waste. Landfill gas doesn't just contain methane and carbon dioxide; it can also contain mercury and other elements toxic to humans and the environment. And landfill liners aren't foolproof. They can—and do—develop leaks.

With old landfills, leakage is a relatively common occurrence. So common, in fact, that in the 1980s the EPA theorized that *all* landfills would eventually leak, as leachate built up and tore through the lining, or as toxic chemicals ate away at it. Remediating these old landfills is highly problematic. In many cases, it requires excavating the landfill, which is a highly toxic task.

Is it possible to make a landfill that won't leak? It's a highly contentious question, one that has led to more than a few heated arguments among the passionate on both sides of the question. Even the best scholars on the topic confess that humans simply don't know.

Considering all of this, if there is to be waste at all, is it still a worthwhile goal to eliminate all landfills?

Many municipalities would say yes and are trying to get to a state called "zero to landfill," where *nothing* gets sent to the trash graveyard. In the United States, the most ambitious city pursuing this strategy is San Francisco. In 2003, the City by the Bay set a goal of eliminating landfill use entirely by the year 2020.

They didn't make it, and pushed the date out to 2030. But they *have* made remarkable progress toward the goal. While the average American diverts about a third of their garbage through recycling and compost, San Franciscans divert 80 percent of theirs, and the number is rising.

But the reduction has come at a cost. The city has legislated mandatory composting, put taxes on certain items, shrunk the size of trash cans, increased fees, employed an array of technologies, and a whole lot more.

In Europe, a company called Contarina has managed to improve on San Francisco's impressive performance. They manage municipal waste for over half a million people in northern Italy and are able to divert over 85 percent of it from landfills, while offering low costs. They do it through extensive education and instituting a system in which everyone has *five* different bins: one for glass, plastic, and cans combined, then one each for yard materials, paper, organic matter, and "other." Different trucks come by, pick up different bins, and take them to different places.

As admirable as all this reduction sounds, is it worth all the effort from an environmental standpoint?

At one level, landfills themselves aren't terribly problematic, and here's why: In your mind, divide up everything that goes into landfills into two buckets. One is the stuff that doesn't decompose on any timeline humans can fathom (and which, in theory, can be recycled or repurposed). We're talking metal, plastic, glass, et cetera. In the other bucket is everything else, from electronic waste to biomass (food and paper) to hazardous chemicals.

That second bucket certainly presents landfilling problems. You don't want to just dump all that stuff in a pit and call it a day.

But what about the first bucket? It turns out you *can* landfill that with very little environmental harm. Those materials don't degrade, and they more or less stay where you put them. According to the EPA, the emissions from landfilling these materials are dramatically lower than what would be required to sort, transport, and reuse most of them. Think about that. More greenhouse gases are emitted dealing with the recyclables than would be emitted by landfilling them.

Further, when you divert materials from landfills, they are never permanently diverted. You can remake those water bottles into raincoats, those raincoats into park benches, and so on. But sooner or later, they will end up in a landfill or somewhere worse. Diverting is only temporary.

"But," you may be thinking, "landfilling those items is still a problem. It would still be better to recycle all that plastic, glass, and metal, and use it again, saving the virgin resources and the energy used to extract them."

But is that, in fact, the case? If you recycle something, are fewer virgin resources actually used?

Intuitively, of course, the answer is yes. However, in their persuasive papers "Material Recycling and the Myth of Landfill Diversion" and "Circular Economy Rebound," Dr. Trevor Zink and Dr. Roland Geyer argue that with respect to recycling, our intuition is wrong. For instance, recycling a pound of aluminum, the most easily recycled material, *doesn't mean a pound less gets mined and refined.*

We spoke with Zink about this counterintuitive conclusion. He is a passionate environmentalist, and he was wary of even speaking to us, because he has seen his views twisted by people to malign environmentalism. But Zink is interested in what really, truly works, and he and his coauthor don't think that recycling comes anywhere close to displacing original production on a pound-for-pound basis.

What really happens, they argue, is that by bringing all of these recycled materials onto the market, the increase in supply forces prices down, and virgin producers are forced to cut their prices as well in order to compete. These lower prices, in turn, *spur additional demand*—and people use more of the material than they would otherwise.

The theory makes sense. As you'll see, when it comes to aluminum cans, if secondary production weren't possible, the cost of an individual can would surely be higher, resulting in a lower consumption of cans.

This phenomenon is more than an interesting academic theory. Zink and Geyer have gone to great pains to quantify exactly how much virgin production is offset by recycling. It varies widely by substance, but to revisit aluminum as an example, recycling 10 pounds of aluminum cans only results in 1 less pound of primary aluminum being produced. Steel is better. If we recycle 10 pounds of steel, 6 pounds less of the metal is made from scratch.

Will we one day mine landfills for all the substances that they contain? It is already done on a limited basis in a few places. Magnets are used to pull ferrous materials out, while conveyer-belt systems can aid in the sorting of other materials. But by and large, this is a difficult task, and raking up all the muck that has been locked away is a substantial environmental hazard of its own. Still, it's more than conceivable that robots in the future will sort through previously landfilled trash in a manner just like the title character in the Disney film *WALL-E*.

One Word: Plastics

As ubiquitous as plastics have become in our everyday life, it's easy to forget that they are quite new. The first plastic made from synthetic components is barely over a century old. Its commercial name is Bakelite, and its chemical name is polyoxybenzylmethylenglycolanhydride, a fact included here mainly to provide a challenge to whoever ends up recording the audio version of this book.

For human use, plastics filled a manufacturing niche that had been filled by natural substances such as ivory, leather, wood, bone, horn, and metals. But those substances were expensive, in limited supply, of inconsistent quality, difficult to work with, hard to refine, or all of the above.

Plastics, on the other hand, were cheap and could be formed into nearly any shape—in fact, the word we use for them originated in a Greek word that means "able to be molded." In a sense, plastics gave us the modern world, and they touch our lives almost every minute of every day.

But if you ever watch movies or TV shows from the black-and-white era, try to spot *any* plastic in any of the scenes, apart from perhaps a telephone. As late as the 1967 movie *The Gradu-ate,* plastics were being touted as the substance of the *future*— one pop culture prediction that has, indeed, come to pass. Today plastics abound in every home, car, and office, and just about everywhere else. Odds are that no matter where you are right now, you can reach your hand out and touch multiple objects made of plastic. If you have on any clothing made with polyester, you're wearing plastic right now.

While millions of products are made from plastic, let's look at just one tiny example: the bread clip. You know the one: the hard plastic closure that seals the (likely also plastic) bag that holds your loaf of bread. Incredibly, almost all of those clips are made by one single company operated by one family. They make billions—yes, billions of these clips every year. And they serve a great purpose, right? Think of all of the waste they've prevented; how much bread has been kept fresh by that one simple device. And that's just one tiny plastic product.

Plastics have many virtues, not least of which is their durability. Some can take up to a thousand years to degrade. However, this durability is a double-edged sword: The same properties that make plastics attractive to manufacturers and consumers also make them difficult to dispose of.

Most of the plastic that we use in the modern world is made from fossil fuels. While many of us may understand that plastic begins as petroleum, how many of us have any idea how a clear water bottle can emerge from the sticky black crude oil or ethereal invisible natural gas from which it is made? How is this alchemy accomplished? Let's walk through it.

Plastics are polymers—long molecules made up of multiple simpler chemical units. The molecules that make up most of the plastic we use today are made up of hydrogen and carbon— hydrocarbons. Hydrogen and oxygen are key components of life on earth, and thus they can be found in abundance in fossil fuels.

In a sense, although we don't think of plastics as "natural" the same way as we would the substances they replaced (e.g., ivory for billiard balls), both have biological origins.

Plastic is made by taking substances such as hydrocarbons, breaking them down into small molecules called monomers, then adding a catalyst that causes these monomers to become attached to each other, forming the polymer we call plastic. Plastic doesn't *have* to be made with fossil fuels, though. You can quite easily make bioplastic at home on your stovetop with just tapioca starch, glycerin, and vinegar. You can even make it with milk and vinegar. In fact, when you "season" a cast-iron skillet with oil over extreme heat, you are actually making a bioplastic coating on the surface of the pan.

Synthetic plastics generally don't biodegrade into other substances. When they reach the end of their useful life, not only are they likely to become waste; they're likely to become waste that sticks around for a long time. Plastics do, however, *photo*degrade, which means that if they are exposed to light, they break down . . . into ever smaller pieces of plastic. These minuscule pieces of plastic enter the food chain, all the way down to plankton. These small creatures are in turn consumed, plastic intact, and so the plastic makes its way back up the food chain to us and other apex predators.

There are plastics that truly biodegrade, or break down into their constituent elements—carbon, oxygen, and hydrogen (in the form of carbon dioxide and water)—and reenter the world in a safe and sustainable way. These biodegradable plastics come in two forms.

The first is bioplastics, plastics made of biomass such as agricultural waste, corn, sugarcane. But just because they're bioplastics doesn't automatically mean they biodegrade; they have to be specifically engineered to do so. At present, those plastics are roughly triple the price of hydrocarbon-based plastics.

In addition to being more expensive than synthetic plastics, bioplastics are less versatile, covering a much narrower range of uses. But both of these limits, price and versatility, reflect that the bioplastic industry is new and small. There's every reason to believe that biodegradable bioplastics could eventually be competitive on cost and function to synthetic plastics. However, our current technological path may not get us there anytime soon. For every 500 pounds of synthetic plastic that is made, only a single pound of bioplastic is produced.

The second way to get biodegradable plastic is to manufacture synthetic, petroleum-based plastics in such a way that they biodegrade. While doing so is technologically possible, there are massive challenges with this approach relating to function, cost, and negative externalities; that is, a significant amount of the costs that arise from these challenges are borne by neither the producers nor consumers of these plastics.

Further, just because something *can* biodegrade doesn't mean it *will*—conditions have to be right. It's also theoretically possible to make compostable plastic, which is different from biodegradable plastic, but to ultimately be effective at reducing waste, that plastic would have to wind up in a compost pile—and even things that people understand are compostable rarely actually get composted.

So, what of recycling? Why not just reuse plastic over and over like we do with steel or aluminum? It's not that simple. "Plastic" is a helpful generalized term for us to use, but polyoxybenzyl-whatever is different from polyvinyl chloride, which is different from polyester . . . you get the drift. Different catalysts and additives give various types of plastic their unique properties. One formulation yields shrink wrap, another squeezable bottles, and another PVC pipe for your plumbing.

To help consumers and recyclers, a taxonomy has been created for plastics that uses a number system. It consists of seven different variants or types of plastics: (1) polyethylene terephthalate (PET), (2) high-density polyethylene, (3) polyvinyl chloride, (4) low-density polyethylene, (5) polypropylene, (6) polystyrene, and (7) miscellaneous plastics. Some believe that the lower the number, the more recyclable, but not so. The numbers themselves don't represent anything in particular.

Each of these types of plastic requires a different process for recycling, and mixing the numbers just a bit can ruin a batch. But that isn't all. The color of plastic has a big influence on whether it's recycled or not. Clear is the best, because it can be more easily recycled into more clear plastic. White is generally okay, because it can be dyed. But nearly everything else, especially the dark stuff, is often landfilled by recycling plants as commercially unusable. There are, of course, efforts to reduce waste through technological innovation. Other initiatives endeavor to repurpose unwanted plastic as insulation, bricks, and playground equipment.

Even interior designers have taken up the challenge. Emeco, the American furniture company that first produced its iconic-shaped Navy chair out of aluminum in 1944, partnered with the Coca-Cola company to produce an identically shaped chair made from the PET (#1) plastic obtained from 111 recycled plastic Coke bottles. The recycled version sells at retail for a little more than half of the price of the aluminum version.

Here's a crazy thought—would it be less wasteful to simply burn used plastic? Most of it is made of fossil fuels, after all, and we routinely burn those for electricity. Plastic is actually more energy-dense than coal, the main fuel for our power grid. Combusting plastic to create electricity is actually quite common: 15 percent of plastic is currently burned as fuel, substantially more than the 10 percent that presently gets recycled. Of course, burning plastic creates its own batch of problems by producing dioxin and acid gases as waste. While today's high-tech plants in places with rigorous environmental regulations are able to capture these byproducts, doing so has a substantial cost.

From a waste perspective, there are other ways to turn used plastic into fuel that aren't quite so problematic. Gasification melts plastics at high temperatures in an oxygen-free environment, which keeps many of the toxins from forming. And pyrolysis allows for shredded plastic to be broken down into diesel fuel. But both of these technologies have critics and challenges, so most of the raw plastic that would be used in these processes just gets landfilled instead.

So that's the problem with plastic waste: Plastics are easy to make, but hard to recycle or convert into usable energy. They are, however, easy to dump into a hole in the ground or into rivers that run to the ocean.

We continue to make three times more plastic every year than we either burn or recycle. It simply accumulates, in an ever-growing pile. So far, in the short time we've had plastics, the world has made enough to bury Manhattan under 2 miles of the stuff. We make new plastic at such a rate—a million tons a day—that it's hard to even get our minds around. Who among us can picture a million tons? And again, that's what we make *every day*.

Think of it this way: Humans make the weight of the Empire State Building in plastics every eight hours. Or, to bring it closer to home, we make the weight of an average house *every four seconds*. Day after day, year after year, it just stacks up.

Further, the production rate for plastics is actually increasing. The authors of a paper titled "Production, Use and Fate of All Plastics Ever Made" concluded that half of all plastic ever created was made in the last thirteen years. When it comes to humans, as William Cullen Bryant wrote, "all that tread the globe are but a handful to the tribes that slumber in its bosom." But when it comes to plastics, the reverse is true—if the trend continues, we'll be making even more than we've ever burned, landfilled, or let escape into the oceans.

How do we get out of this cycle of waste? There are at least three obvious paths out. First is to change the behavior of humans and how we deal with plastics. We can do so a number of ways, including legislation, public awareness, deposit schemes, and more, all of which are explored elsewhere in this book.

The second path out is to change the plastic itself through technology. Cheap, versatile plastics can be made to safely biodegrade under normal conditions, but we aren't there yet economically and probably won't be anytime soon.

Third, as plastics themselves were a replacement for wood, tin, horn, and bone, it's possible that humans may discover a way to produce other substances that have properties superior to plastics.

But until we go down one of these paths, plastics will keep stacking up.

They're Everywhere, Man!

As addressed in the previous chapter, we produce plastics at three times the rate we dispose of them through recycling or burning. So the waste just piles up, ad infinitum. From plastic bags to plastic straws to six-pack rings to microbeads in toiletries, the impact of plastic waste is a topic as ubiquitous as the plastic itself.

A lot of plastic ends up in landfills, which isn't the worst place for it, as we explored in the chapter on landfills. Other bits end up in the oceans, which is a far worse place. Some plastics photodegrade or are broken down mechanically into microplastics and end up in the air and the soil. And a great volume is metaphorically swept under the rug, left to rot in fields or be piled up in a heap somewhere.

The amount of plastic that goes into oceans is staggering: 500 pounds of waste plastic finds its way there *every second*. Until recently, though, we barely bothered to give it a second thought. After all, as eminent natural historian Sir David Attenborough explains, "for years we thought the oceans were so vast and the inhabitants so infinitely numerous that nothing we could do could have an effect upon them. But now we know that was wrong."

If we dump 500 pounds of plastic a second into the ocean, that's about 8 million tons a year. Compared to the 400 million tons of plastics produced each year, this quantity may not seem like a great deal. But once plastic makes it into the ocean, it pretty much just . . . builds up. There is effectively nothing downstream from the oceans. The World Economic Forum famously estimated that by 2050, the weight of all of the plastic in the ocean will exceed the weight of all the fish, and while that assertion may not technically be true—we have no way of credibly estimating the weight of all the fish in the sea, so the WEF's estimate was more political than scientific—the fact that the comparison is even roughly true merits real concern.

But how does plastic get into the oceans in the first place? Each year, 20 percent comes from ships. While this figure includes waste dumped by cargo ships, naval vessels, cruise liners, and pleasure craft, the greatest source is fishing boats discarding nets and traps.

Another 30 percent comes in via rivers when they flow to the sea. While there are 165 major rivers on earth, only ten—eight in Asia and two in Africa—account for virtually all of the plastics that make their way from river to ocean. One single river, the Yangtze in China, contributes 1.5 million tons annually to the oceans: That's more than 100 pounds a second.

Two sources, ships and rivers, account for half of the ocean's plastics. The other half comes from a hodgepodge of sources, including illegal dumping, windborne trash, plastic that escapes landfills, and materials washed in by natural drainage or storm sewers.

How concentrated are plastics in the ocean? That's a complex question. You have probably heard of the infamous Great Pacific

Garbage Patch, the Texas-sized region of the Pacific full of plastic trash. Articles written about it often use stock photography showing water densely packed with garbage. However, these visualizations aren't good representations of the patch for several reasons. First, half the weight of the plastic consists of one single item: fishing nets. Second, at its densest, the patch contains about 1,000 pounds of floating plastic per square mile. You could be in the middle of it and not see any plastic at all. In fact, the entire weight of the plastic in the patch is "only" 80,000 tons, the amount dumped into the oceans every four days. The plastics in the ocean aren't concentrated in a convenient enough place to make cleaning them up easy. Some plastics have found their way to every corner of the ocean, even down to the deepest spot in the oceans, the bottom of the Mariana Trench.

Altogether, there are now about 160 million tons of plastic in the oceans. While plastics can photodegrade, they only really do so effectively when they are dry, and so the earth's oceans, which are famously wet, hamper such degradation. Even when plastics do break down, they either release toxins or remain chemically unchanged but instead become microscopic in size, entering the food chain.

Plastics in the ocean, however widely distributed, cause a good deal of harm. They kill marine life in a number of horrific ways. In some parts of the world, virtually all the sea birds have plastic in their digestive systems, and those same seabirds feed plastic to their chicks. Plastic ingestion has become the number one cause of death among the chicks, killing up to half of them in some species. In addition, the plastic itself can absorb organic pollutants from the ocean's water, making it even more harmful to sea life when ingested.

Beyond the oceans, plastic finds its way into the soil and air as well. The authors of a paper called "Atmospheric Transport and

Deposition of Microplastics in a Remote Mountain Catchment" have found microplastics in the most unlikely of places: the remote, pristine peaks of the French Pyrenees. Researchers were curious why microplastics the size of viruses and large grains of sand were found in soil and water samples from such a sparsely populated and little-industrialized part of the world. Using atmospheric equipment, they began looking up in the air for these particles and found them in abundance. The best models suggest these particles blew in from at least 70 miles away and perhaps much farther. This dusting of the surface of the planet with plastic might be of use to archaeologists in the distant future, in the same way that the highly concentrated layer of iridium deposited by the meteor that wiped out the dinosaurs is used to demarcate the boundary between the Cretaceous and Paleogene eras.

According to a paper called "Human Consumption of Microplastics," Americans inhale or consume an average of 100,000 pieces of plastic per year, although the number varies dramatically by person. The World Health Organization estimates that bottled water contains, on average, over a thousand pieces of plastic per gallon. Of the hundreds of bottles of water they examined, well over 90 percent had some amount of plastic floating in the water.

Should we be worried that we consume so much plastic in so many different ways? After all, the list of foreign substances we ingest in our food each year is pretty, well, unappetizing. The FDA allows 1 ounce of cinnamon to contain 200 insect parts, and it permits measurable quantities of mold and rodent droppings in other foods. A 20-ounce can of tomatoes is allowed to have one entire maggot in it. But whether it's maggots or plastic, what matters in the long term is the amount of these items we consume. The sixteenth-century Swiss physician Paracelsus quipped, "Sola dosis facit venenum"—the dose makes the poison. In other words, everything is poison at high enough levels, and nothing is poisonous at low enough ones. What of plastic consumption and humans? Is 100,000 particles a year dangerous?

Intuitively, one would think so. However, a number of distinguished academics in environmental science reject the notion that the levels of microplastics we encounter in the environment are severe enough to be harmful to humans or marine life. Consider Professor Allen Burton of the University of Michigan, who has a PhD in aquatic toxicology and is the editor in chief of one of the premier journals for environmental toxicology. He wrote a paper bemoaning the fact that scientists are propagating this idea: "I find the continuing publication of microplastics studies stating a severe environmental threat, in high quality journals disturbing. These studies are rapidly picked up by the news media, as we have seen and serve to misinform the public and policy makers, as noted by others."

Burton and his colleagues are right that we shouldn't jump to such conclusions, but the truth is we really don't know. Many believe the plastic we eat just passes through us with absolutely no deleterious effect, and the plastic we inhale pales compared to the other pollutants in the air. Many of these folks worry that a fixation on microplastics is distracting from real environmental and health issues. The European Commission's Science Advice for Policy by European Academies sums up the situation: "The best available evidence suggests that microplastics and nanoplastics do not pose a widespread risk to humans or the environment, except in small pockets. But that evidence is limited, and the situation could change if pollution continues at the current rate."

This is the challenge with plastics. There's just so much we don't know. The oceans are largely opaque to us, and estimates of the amount of microplastics in them vary by orders of magnitude. We aren't sure how plastics are getting everywhere we find them. We aren't sure what the ocean is doing to the plastics dropped into it, and we aren't sure how all of this affects the earth's ecosystems and our lives. The proliferation of plastics is unquestionably worthy of our concern and attention.

The Problem with Single-Use Plastics

Although plastics are made to be highly durable, about half of all plastic made is intended for one-time use, mainly as packaging.

Among single-use plastics, the two most visible and symbolic manifestations are plastic bags and plastic water bottles. There are significant similarities between these two forms of plastic, not least of which is that they are consumed at approximately the same (staggering) rate: about a million of each per minute, every minute of every day. In the United States, we use roughly one plastic drink bottle and one plastic bag per person each day.

Let's examine each of these single-use plastics and explore the challenges in getting to a world where we're able to eliminate the waste associated with plastic bags and bottles.

Younger readers might not appreciate just what a new phenomenon bottled water is. Until the 1980s, the only bottled water most people had ever tasted came from the large multi-gallon bottles atop water coolers in offices—and those bottles were more likely made of glass and were cleaned and reused. In the early days, bottled water was a joke—the idea that a person would pay the price of gasoline for something they could get from a tap virtually for free was viewed as ridiculous. Comedian Jim Gaffigan pokes fun at the whole notion in a special for Comedy Central that starts out: "How did we get to the point where we're paying for bottled water? That must have been some weird marketing meeting over in France. Some French guy's sitting there, like, 'How dumb do I think the Americans are? I bet you we could sell those idiots water.'"

But today, we are in a completely different world. Bottled water is so ubiquitous that new sports stadiums have been built without any drinking fountains at all, in order to provide opportunities to sell water for as much as $5 a bottle.

Municipal water, one of the crown jewels of U.S. infrastructure, is often regarded as being of inferior quality to bottled water even though, apart from a few notable and newsworthy exceptions, that's just not true. John Jewell, writing for *The Week,* characterized bottled water as "the marketing trick of the century."

It's not that *water* bottles themselves are a big deal; it's single-use plastic beverage bottles in general. One company alone, Coca-Cola, is estimated to be responsible for 20 percent of all the plastic bottles in the world, and the majority of those are not filled with pure water.

While most plastic bottles can be recycled, as we've seen in other chapters, recycling rates vary widely around the world. In the United States, about a quarter of bottles are recycled. A 25 percent rate is actually quite high compared to the American average for plastic recycling in general, because most everyone knows bottles can be recycled. Recycling bins often have a plastic drinking bottle printed on them as part of their iconography. But the 25 percent rate in the United States pales in comparison to other parts of the world, including the overachieving Norwegians, who recycle a staggering 97 percent of plastic bottles.

In fact, Norwegian recycling has become so advanced and efficient that it's economical to remake clear plastic bottles into new clear plastic bottles, time and time again. How do they achieve that outcome when other countries can't?

Through two different financial incentives.

The first relates to *producers* of plastic bottles, whom the Norwegians tax. But the tax tapers down to zero depending on how the nation collectively recycles. If Norway hits a 95 percent national recycling rate, there's no tax at all. For about the last decade, they have achieved this benchmark every year.

The second incentive relates to *consumers* of plastic bottles. Norwegians pay a relatively large deposit—between a dime and a quarter—for each plastic bottle they buy, which is then refunded when the bottle is turned in to a recycling center. There are many such centers and they're fully automated, making it easy and worthwhile for Norwegians to do the work to get bottles into the recycling stream.

Another key to Norwegian success is not vilifying plastic bottles themselves. Sten Nerland, director of logistics and operations for Infinitum, which oversees this program in Norway, explains, "As an environmental company you might think we should try to avoid plastic, but if you treat it efficiently and recycle it, plastic is one of the best products to use: light, malleable, and it's cheap."

Germany has a similar system and achieves similar results. In Germany, the 25¢ deposit is paid to the manufacturer of the plastic item; they get a bit of profit from the 2 percent of bottles for which they collect a deposit but will never have to pay out.

As a result, a substantially higher percentage of bottles are now made out of plastic rather than reusable materials such as glass, mainly because Germany has a predictable and developed infrastructure in place to handle the deposits, collections, and refunds.

In both systems, while average citizens are conscious of the desirability of recycling, the very high rates they achieve are due in part to a population of financially disadvantaged citizens who make a meager living going through trash finding tossed plastic bottles. To some this practice is viewed as acceptable, culturally speaking—it can be seen as a tax on the wealthy that benefits the poor, as well as a mechanism to provide employment to an otherwise underemployed group of people—but it challenges the notion that *everyone* in these countries has bought into the system of collective recycling.

In contrast to plastic bottles, plastic bags have one of the lowest recycling rates, at around 1 percent. Why so low? A few reasons. First, they are plastic film—think of the plastic bag covering your dry cleaning or a newspaper on a rainy day—and need to be recycled using a process different from that for water bottles. Second, in most places they aren't even allowed in recycling bins since recycling centers don't want them.

But why? Partly, it's because bags resist being sorted on conveyor belts; the smallest breeze is capable of moving them from one part of a recycling plant to another, and they can jam up machinery. One plant reports that the plastic bags people (incorrectly) put in their recycle bins shut *everything* down *every couple of hours*. The city of San Jose states that it spends more than a million dollars a year to repair recycling equipment that breaks as a result of plastic bags. To recycling centers, plastic bags are a major nuisance, the waste world's equivalent of stepping on bubble gum.

Single-use plastic bags are particularly reviled because their duty cycle is so short compared to the time it takes them to decompose. The average bag is used for about half an hour, but it persists for another century. They overstay their welcome, so to speak. Further, plastic bags are a highly visible form of litter. For the same reasons they don't stay on conveyor belts, they also don't stay in landfills, and seem to be attracted to trees in much the same way as Charlie Brown's kites are.

While bags don't account for a large percentage of overall litter, they're disproportionately destructive as a result of the impact they have on marine animals. They're frequently ingested by
dolphins, whales, and turtles, who mistake them for jellyfish. In
addition, algae stick to them. As enzymes from the algae cause
the bags to degrade, the process results in odor that makes other
animals think they are food. They harm terrestrial animals as
well, with cows being particularly prone to eating them. Nothing
good can come from eating a plastic bag.

Usage rates for plastic bags vary tremendously by country,
with the average American using 350 a year—nearly one a day—
and the average Dane using just four. Why just four? Their low
usage rate is achieved by taxing the bags and then providing convenient durable, reusable ones in their place. The four that the
average Dane uses probably are to get fish from the fish market or
takeout from a restaurant. They are simply not seen in grocery
stores. As of this writing, Denmark (like many U.S. municipalities) was considering an outright ban on them, along with a ban
on stores giving away non-reusable bags—or at least they were
until about March 2020.

The trend away from single-use bags—and single-use plastics
of all kinds, actually—came to a screeching halt with the COVID
pandemic. No longer could consumers take in their cotton bags
(especially if they had switched to grocery delivery). Several
states with laws on the books that discouraged or even banned
single-use plastic bags have suspended them. California was one,
and credible estimates peg the increase in the number of plastic
bags consumed at half a billion—just in that state. Some are concerned that even when the measures are reinstated, consumer
behavior will have changed so much that it will be difficult to get
back on track.

Pre-COVID, bags were largely taxed in Europe and banned in
dozens of countries around the world including India, France,

and Italy. In China, *free* bags were banned. Much of Africa bans them, including Kenya, where half of all cows were found to have a plastic bag in their stomach. That country boasts the most draconian laws, where making or selling a plastic bag can result in a $40,000 fine and four years in jail. Even using a bag can set you back $500 and land you in prison for a year. According to the UN, two-thirds of countries have some kind of national legislation on plastic bags.

Absent the effects of a pandemic, bag policies had been around long enough that we can now see enough data to determine the difference in efficiency between taxes and bans. Interestingly, taxing bag use by even a seemingly modest amount, like a dime, results in nearly a 90 percent reduction in use—very close to the reductions that result from outright bans. Even a 5¢ tax in Washington, D.C., achieved a similar result.

Bans are generally favored by environmentalists, but consumers are split. Some want to retain the choice to use single-use bags in particular situations, while others view the taxes on bags as, well, an increase in their taxes.

Plastic companies, of course, hate both bans and taxes. And retailers are generally against them when they're imposed on a city-by-city basis, fearing customers will change their habits to shop in places without a ban or a tax. Moreover, these taxes are considered regressive, since both the poor and Bill Gates pay the same amount per bag.

Compared to plastic bottles, deposit schemes for bags are nearly unheard of for fairly obvious logistical reasons. Plastic bottles can be bar-coded and easily returned, but bags resist such orderly handling. However, a tax on bags and a deposit system on bottles function effectively the same way—they increase the cost of the item to the consumer, causing consumers to use the items (and subsequently discard them) less.

So how should we treat plastic bags and bottles as we strive to curb our waste? Would you ban both? Tax them? Have a deposit system? At the extreme, we could outlaw single-use plastics and institute the death penalty for their use. Usage would go very close to zero. Likewise, we could impose a $100 per bag or bottle deposit system or a $100 tax. Again, usage would get close to zero. Or we could revert to the previous status quo, where there are no restrictions at all. How do you navigate these choices? Which strategy is least wasteful?

Determining an optimal strategy is fairly straightforward, at least in theory. First, we would need to quantify the total cost of all the bags: The dead cows and sea turtles. The urban blight of bags in trees. The cost to transport them as waste to landfills, and the environmental consequences of doing so. The costs of producing them and transporting them in the first place that aren't captured in the price of the bags themselves—for example, the externalities in production and transportation—and so on.

Let's say we determine that the 500 billion bags used each year inflict a $500 billion aggregate cost on the world. That means every bag imposes $1 of external costs to the world, and should therefore be taxed at $1. If a bag is worth more to you than the $1 in damages it inflicts on the world, then you will pay the dollar. If it isn't, you won't.

Would a deposit of $1 give the same results? Not exactly. If we assume that the cost to society of a bottle thrown in the trash (or discarded as litter) is $1, but the cost to society of a bottle returned to a recycling center is zero, then yes, it would be the same. Because in this case a person will either (1) pay a $1 deposit, use a bag, throw it away, and inflict $1 in damage to the world or (2) pay a $1 deposit, use a bag, return it to a recycling center, have no impact on society, and get their dollar back.

But the reality is that the bottle returned to the recycling center still inflicts a cost on society—another example of those production externalities we talked about earlier.

In this case, a deposit doesn't actually reflect the bottle's cost to society. It might if you paid $1 as a deposit but only got 90¢ back—but then again, that 10 percent effectively becomes a tax.

The German example given earlier in this chapter nicely illustrates this principle. Because almost all of the deposits on bottles were refunded, production wasn't disincentivized—there was no disincentive to *make* bottles. Rather, people were disincentivized to *throw bottles away* rather than recycle them. Production thus was higher than it would have been otherwise.

Bans probably provide the least efficient outcome. While they may cost less to enact and enforce, at least in countries with widespread compliance, the implicit assumption is that the cost of a bag on the environment is infinite. In other words, even if a bag is worth tens or hundreds of dollars to you at that moment, you could not be sold one. Bans work best when a society concludes that the thing it wants to ban has no uses that benefit that society—as we've done with meth and hand grenades, for example. Plastic bags don't fall into the same category.

So deciding how to curb this waste requires knowledge of how much the *creation* of an item damages society and how much the *disposal* of that item damages society. With those two pieces of data, you can determine if a tax, a ban, or a deposit scheme is the best choice.

You have to do the same for the substitutes as well—whether we're talking about reusable cotton grocery bags or glass jars. And in this case, those substitutes may have significantly higher costs than the economic costs of bag creation and disposal. A 2013 study by academics at the University of Pennsylvania and George Mason University suggested that emergency room visits from foodborne illnesses increased in San Francisco after that city passed its plastic bag ban. And the 2020 experience with COVID has caused a global reassessment of the net value of reusable plastics—a critical component in determining the extent to which they are waste.

Another study by a University of Sydney researcher found that when cities in California banned single-use plastic bags, consumers indeed used many fewer shopping bags, leading to a reduction of 40 million pounds of waste. However, at the same time, sales of small trash bags rose in those places, resulting in a net reduction of only 28 million pounds, as consumers presumably substituted "official" waste bags for their repurposed alternative.

If it were possible to know all of the ripple effects of these policies, we could minimize waste. But it's unlikely we'll ever have that level of insight.

Even if we knew all of the results, quantifying each would be virtually impossible. How much is a sea turtle worth? What's the value of trees that are free of plastic bags? These become value judgments in the end, and widespread consensus has not yet been reached.

The Aluminum Cans . . . and Cannots

As you'll see in a later chapter, there is a tremendous amount of waste involved in mining the aluminum we use for our soda and beer cans. Moving thousands of tons of earth to find bauxite ore is, to say the least, an inefficient process, and the amount of energy required to convert ore into usable aluminum metal is staggering.

Is recycling aluminum less wasteful, though? Creating new aluminum from already existing sources of the metal—recycling—is referred to as secondary production, as opposed to primary production, which is mining it from the earth. In the United States, 60 percent of aluminum is primary production and 40 percent is secondary.

When it comes to metals such as aluminum, copper, and steel—in contrast to many other materials, as we examine elsewhere—recycling is a more efficient way of producing usable supplies of each metal. Because aluminum in particular can be recycled quickly, inexpensively, and virtually an infinite number of times, if you drink a soda today and throw the can in your recycling bin, odds are someone else will be drinking from that same aluminum in just two months.

How much more efficient is secondary production of aluminum? In terms of electricity use alone, recycling used aluminum into fresh aluminum takes just one-fifteenth of the electricity used in primary production. So instead of the 15,000 kilowatt-hours (kWh) of electricity consumed in primary production, recycling that same ton requires only 1,000 kWh.

Given the reduction in electricity use, recycling plants need not be located in places where energy is dirt cheap. In fact, though we have only a sprinkling of primary aluminum smelting plants in North America, there are more than a hundred secondary aluminum-producing locations spread across thirty-one states and two Canadian provinces.

No matter where your recycling bin is, it's unlikely that your can is going to travel more than a few hundred miles to where it can be melted down and used again. Compared to the 3,000 miles the raw materials for primary production travel, that's a lot of waste eliminated.

Apart from the transit issues, what other advantages does recycling have? One is a reduction in fossil fuel consumption. The world still generates the majority of its electricity from fossil fuels—and of those fossil fuels, coal is by far the largest source, accounting for nearly 40 percent of production. It takes roughly a pound of coal to make one kWh of electricity. So for every ton of aluminum recycled, we're able to avoid burning 14,000 pounds of coal. (This assumes that recycling a ton of aluminum means a ton less is mined, which isn't necessarily the case, as we discussed in the chapter on landfills.)

What's more, by preventing the combustion of those 14,000 pounds of coal, we prevent the release of *40,000 pounds* of carbon dioxide. That isn't a typo. In combustion, carbon combines with atmospheric oxygen to produce more CO_2 than the weight of the coal burned. Saving that much CO_2 is the equivalent of planting 210 trees and letting them grow for ten years.

This analysis assumes, for simplicity's sake, that the electricity used to make both primary and secondary aluminum is from coal, which isn't really the case. Primary production usually uses more green energy, since refineries tend to be built to be near sources of cheap energy. So the greenhouse gas emissions of primary production are around seven times greater than those associated with secondary production (as opposed to fifteen times greater, as their relative energy consumption might suggest).

Given the tremendous reduction in waste from recycling aluminum, why isn't *all* aluminum production secondary? Haven't we mined enough that we can just use the same aluminum over and over?

As it turns out, we haven't. Demand for aluminum is increasing by about 3 million tons a year and is expected to continue climbing. Much of our use of aluminum is for the long term, such as in infrastructure or residential and commercial construction. So we constantly need new primary production.

The COVID epidemic had an interesting—and in some places counterintuitive—impact on the can industry. As bars and restaurants closed, the demand for canned drinks of all kinds went way up. That's because fountain drinks and keg beer were replaced at home with single-serving containers. Meanwhile, recycling capacity went down, as many facilities were shorthanded. The result was a can shortage, or more particularly an aluminum shortage. U.S. manufacturers were buying every bit of aluminum they could lay their hands on. They turned in particular to Mexico and Brazil, which suddenly had surpluses, since in those countries bars and restaurants typically served canned drinks, and when that demand went away, the total sale of canned goods plummeted in those places.

What would it look like if an individual decided to get into the secondary aluminum production business? In a later chapter, we'll address the issue of whether basement smelting of bauxite can be economical, and won't give you any hints at this point—though your instincts might be leading toward a preliminary answer. But could secondary production of aluminum be a cottage industry? What would it take for the average person to create a ton of secondary-production aluminum?

First, you would need to acquire the raw material: recyclable aluminum. Aluminum is used for a wide variety of purposes. About a tenth of all aluminum is used to build electrical equipment. A fifth is used in construction, a fifth in transportation, and a fifth in food and beverage packaging. It's that last one we think of most often when it comes to recycling—namely, aluminum cans.

Given that it takes more than sixty thousand aluminum cans to produce a single ton of aluminum, you aren't going to drink your way there anytime soon. You would have to buy the aluminum.

Thirty-two 12-ounce cans make up a pound. Recycling plants buy a pound of cans for an average of about 45¢. If you could match their price, you could buy a ton of used aluminum for about $900. At residential rates, the electricity needed to process it would cost about $100. And since recycled aluminum is chemically identical to primary production aluminum, it sells for about the same price: $1,500 a ton or so.

So, if we ignore the fixed costs of a home Smelt-O-Matic, it would appear as if basement smelting could replace Amway or Mary Kay as a profitable home business. However, as is usually the case, it's not so simple.

For one thing, there's a plethora of other costs involved along the way. And for another, you're probably not going to be able to unload a ton of aluminum at your next garage sale. Plus operating at a smaller scale than industrial recycling operations means that you'll be more sensitive to changes in input prices of raw material, and that other costs will eat into your margins and make your home business less profitable than a major operation would be. (As of this writing, for instance, the online legal services company Rocket Lawyer didn't provide any do-it-yourself options for drafting commodities delivery contracts, or for applying for the permits you will undoubtedly need to operate this operation legally.)

Given all the factors that affect the price of finished aluminum—everything from energy subsidies in China to tariffs in India—you're probably better off sticking to your day job.

Recycling the Truth

What happens to the objects that are recycled? When you toss a water bottle into the blue bin, where does it go? When you recycle old office paper, what mechanisms allow those wood fibers to be used elsewhere? How does it all happen, and what are the fundamental economics that drive the industry? The answer might surprise you.

Let's start with how it's *supposed* to work.

In the most common scenario, we begin by sorting trash and recyclables into two different containers. Recyclables head to a plant where they are further sorted by type—the different kinds of plastics, paper, cardboard, glass—then baled. Each material is sold to a different recycler for a different price, and the money that is generated ostensibly funds the whole operation. All of this effort requires energy, transportation, infrastructure, labor, and the like, but we hope that the inherent value of the materials themselves will pay for the overhead.

If we adopt a "do no harm" approach—a recycling Hippocratic oath, so to speak—at the very least we assume that our actions don't actually result in a net increase in waste and pollution compared to landfilling the refuse and replacing the recycled goods with primary production. Instead, we're trying to create a "cradle-to-cradle" system, whereby the waste from one process or product becomes the raw material for another one—an ecosystem that's less wasteful than starting from scratch.

That's the theory. This halcyon portrait of a circular economy in action—which the very logo for recycling itself evokes—is what we might expect to see painted by a green Norman Rockwell. The reality is very different. For certain materials, recycling works well. For instance, much industrial waste is homogenous and easy to recycle at a metal stamping facility or a paper mill. Scrap iron recycling is a vigorous business without much waste.

But the bulk of what we picture as the recycling process has a less rosy reality.

Why is that? Aren't plastics, paper, and glass in high demand? Don't recyclables have significant economic value?

The problem, as it is with so much of the waste we generate, is with human behavior. A full quarter of recyclable materials arrive at their destination contaminated in one way or another. That's on us. According to Waste Management, when we put something in a recycling bin, we get it wrong one time in four. That's a pretty high error rate.

What drives these mistakes? First is "aspirational recycling." Most of us mean well—so much so that we feel virtuous dropping something into a recycling bin instead of the trash. But that tiny endorphin rush results in waste when we're not careful.

Are clothes hangers recyclable? Who knows? Better put them into the recycling bin just in case. Plastic bottle caps? Juice boxes? Water hoses? Christmas lights? Egg cartons? None of these, by the way, are usually recyclable. But all frequently wind up in the bin.

Making matters worse, what can and can't be recycled varies

widely by place and time. What's recyclable at the office might not be recyclable at home, or at a hotel on vacation.

Additionally, while products might be made entirely of materials that individually are recyclable on their own, the finished products themselves may not be. Think of a wax-covered cardboard milk carton with a plastic pouring spout. It's neither fish nor fowl, and it actually belongs in a landfill or an incinerator. But often it gets put into the recycling bin since it sure *looks* like it should be recyclable.

Some argue that an issue in the United States is the practice of single-stream recycling, which is uncommon in other parts of the world. Single stream means you get one recycling bin and all your recyclables go in it, to be sorted later. As a process, single-stream recycling often causes waste. If a newspaper and a glass beverage bottle are in the same bin and the bottle shatters as it's dumped into a collection truck, it can ruin all the paper. Plastic bottles that still have water in them can turn recyclable cardboard into useless mush. Undoubtedly, going to dual or triple stream would reduce the contamination problem by shifting the responsibility for sorting to the general public—at the risk of lower compliance.

The advantage of single stream is that it's easier on consumers who, we assume, won't keep up on sorting into and maintaining two, three, or four different recycling bins. Individuals make mistakes, and an argument for single stream is that it's better for the experts to sort it all out. Also, multi-stream plants assume they're getting uncontaminated streams of recyclable materials, so they aren't set up to sort through every piece. In the single-stream facility, trained employees produce better results.

In theory, we could use more labor at recycling plants to ensure the purity of material. But doing so would add substantial costs to what's already a barely functioning business model, as we will see.

Technology may ultimately lend a hand here. Robots that can use artificial intelligence to recognize and sort items are already deployed in some places, and the technology will only get better, faster, and cheaper. Over the long term, machines will likely do a far better job than humans at sorting. Unfortunately, we're not quite there yet.

It makes sense to ask—has recycling, as the ideal process described at the start of this chapter, *ever* worked?

From one perspective, yes, because we had a shortcut. We could ship our materials to China, where low-wage workers took over the laborious process of carefully sorting all the materials. According to the EPA, 60 million tons of goods travel through U.S. recycling plants per year. Until recently, China took a quarter of that amount. China's importation of trash came about for a pretty interesting, waste-reducing reason. For a long time, container ships full of Chinese manufactured goods would make their way to the United States, then turn around and sail back empty. As a result, the price to ship a container *to* China from the United States was a tiny fraction of the cost going in the other direction. In fact, it was so inexpensive that it made sense to put literal trash in those containers. So we did.

Early on this practice proved to be a boon for some Chinese people. The most notable example is Zhang Yin, who is one of the richest self-made women in the world. She came from a very modest background in China and ended up in the paper business. She found raw materials in China to be both in short supply and of poor quality. This sparked an idea, so in the 1980s she and her husband moved to the United States and drove around in a Dodge Caravan looking for scrap paper, which they found in abundance. They would load their van, drive it to the port, and gradually fill up a shipping container. That container would go to China, where the material would be made into corrugated cardboard boxes, which were sold to Chinese manufacturers, who would then ship those boxes back to the United States full of merchandise to start the cycle over again. Zhang Yin built an empire and amassed billions on the back of this business model.

But the landscape has changed dramatically. In March 2018 China implemented Operation National Sword, designed to all but eliminate *yang laji*, or foreign garbage. These new policies banned the import of many materials and raised the purity requirements on others to 99.5 percent, a standard deemed by many to be virtually impossible to meet. Human sorters, even on the best of days, aren't consistently that accurate. The impact of this policy was dramatic: Over the course of a year, exports of American recyclables to China fell by 90 percent.

Why did China institute these policies, and why so abruptly? According to the Chinese, it was for public health and pollution abatement reasons. There's clearly truth to this assertion, as many of the individual Chinese companies that were importing the garbage weren't handling it responsibly and it was ending up as a blight on the Chinese landscape. Too, the optics were problematic: Being a world superpower doesn't necessarily mesh well with being the world's trash buyer. Finally, the Chinese market has plenty of its own recyclables to deal with. Interestingly, the shifting economics of Chinese recycling prior to 2018 accelerated the demise of the mass-market paperback book business in the United States. Unlike hardcovers, unsold paperback books aren't sold at a discount; they're stripped of their covers and become wastepaper. For decades, as long as the Chinese market for scrap paper was robust, it made sense for retailers to carry a large stock of paperback books. If those books didn't sell, some of their cost could still be recaptured by selling the scrap paper. When the scrap paper market collapsed, however, disposing of unsold paperbacks became a pure liability, which encouraged retailers to use that shelf space for more profitable items.

So where can our recyclables go? Other countries with low

labor costs, including India, have set up similar import restrictions. Some small countries still import recyclables, but none has anywhere near the capacity of China. What are the other choices? That's a problem that has no easy answer, and many believe that trash shouldn't be exported—that if a nation creates a certain amount of trash, it should handle the waste itself.

Operation National Sword was a watershed event. Overnight, the economics fell out from under the recycling industry. Materials that formerly could be sold for $50 to $100 a ton suddenly fetched nearly nothing. In some cases, the economics completely inverted: Instead of getting paid for their bales of recyclables, municipalities now had to pay $50 to $100 a ton to have them removed. Imagine that business: You have trucks that gather recyclable materials, you pay humans to sort it, and then when you have a finished product—a bale of mixed paper, for instance— you have to pay someone to haul it off. You don't have to be Warren Buffett to figure out that's not a business with a future.

It's wrong to say China is the source of all the recycling industry's woes. Cheap oil and natural gas, courtesy of new technologies such as fracking, mean that primary production of plastics is less expensive than it used to be. As a result, your used water bottle, which is essentially made of highly processed fossil fuels, is worth less as well, since it competes with newly produced bottles.

Because of all these economic factors, municipalities are being forced to rethink strategies or abandon recycling programs altogether. In some areas, consumers sort their recyclables out from their garbage, two separate fleets of trucks pick it all up, and then it's remixed together in a landfill. This is not an isolated phenomenon, either. A recent *New York Times* article refers to "the hundreds of towns and cities across the country that have canceled recycling programs, limited the types of material they accepted or agreed to huge price increases."

In some municipalities, such as Philadelphia, the majority of recycling is quietly burned to generate electricity. Such practices often take place without fanfare due to fear that if consumers know their recyclables will end up in a furnace or a landfill, they'll stop recycling, and down the road, when the economics of the industry have righted themselves, consumer behavior will have to be changed again. There are only so many times a town can change what is and isn't recycled without fatiguing or confusing its residents. Better, the reasoning goes, to just keep having residents separate their discards as usual, even if it means landfilling the carefully sorted materials. Other towns are simply ending their now-costly recycling programs.

Again, not all recycling is broken. In non-coastal areas where the shipment of materials to China required extra expense, China's ban on foreign trash had less impact. For glass, which is so heavy that it was never economical to ship, the economic situation isn't dire. Neither is it the case with high-value items such as copper, aluminum, iron, and steel. Additionally, there are regional dynamics in play in parts of the United States where materials are recycled profitably. But overall, the idea that what we put into recycling bins gets profitably recycled most of the time is false.

Economically speaking, recycling has always been on pretty thin ice. The overall economic analysis of recycling seldom measures all of the external factors involved. You can compute the economic and environmental costs of the trucks that gather the recycling, and sometimes people do. But seldom do other factors get included, such as the water used to rinse out recyclables, or the time people spend sorting their trash. When all these factors are taken together, it's unclear that recycling "works" in any significant economic way for many goods.

But does it matter? We can, as a society, decide to regard recycling not primarily as an *economic* activity but as a *moral* one.

Sure, it would be nice if the economics supported widespread recycling. But if they don't, it doesn't matter, because recycling is simply the right thing to do, even if it costs more. If that's our collective decision as a nation, then public policy must clearly replace the illusion of free markets as a driver of the recycling industry. In such a case, garbage should be taxed, recycling subsidized, landfilling recyclable materials outlawed, and refundable deposit schemes enacted for all kinds of materials.

There are parts of the world where high rates of recycling are achieved through these methods—that is, through taxes, subsidies, and social pressure. Taiwan is one such case. Businesses are expected to handle their own waste or subsidize the government's recycling efforts. Consumers have to buy trash bags for non-recyclables, effectively taxing trash. Items that can be recycled, on the other hand, can be disposed of freely. Trucks roam the streets playing music, like ice cream trucks in the United States, signaling to the population that it's time to bring out their recycling, and volunteers are on hand to offer suggestions for how to more effectively recycle. Finally, the government subsidizes innovation to build products made from recycled goods. This last option can have the unintended consequence of lowering the effective cost of those materials, increasing their usage. So policy here must be carefully thought through.

But is there a path forward whereby simple market forces can turn the sow's ear of our trash into silk purses? Can recycling be done profitably on a large scale nationwide?

Markets are powerful social forces, and if new water bottles could be profitably made from used ones instead of virgin plastic, that's a win for everyone. There are many reasons to think this might happen, most of which relate directly to technology.

The innovations are numerous and significant, and come along at an ever-increasing pace. AMP Robotics has built an AI-powered robot system, named Cortex, which they claim can sort recycling with up to 98 percent accuracy; it can even read the bar codes on items to know what they're made of. Scientists out of Texas A&M AgriLife Research have found a way to turn paper waste into high-quality carbon fiber, suitable for manufacturing new items. Reebok is making compostable shoes out of cotton and corn, while LEGO is making some of its pieces out of plant-based plastics. And in Melbourne, Australia, there's a highway made of recycled plastic mixed with asphalt. New techniques are being developed to make recycled plastics competitive with virgin materials in terms of quality, while similar strides are being made with recycled glass. New venture capital funds are springing up to fund innovation in recycling. And the generation of young people growing up now recycles instinctively; a million business plans are gestating in their minds, trying to crack the code of making garbage pay.

Recycling is an incredibly complex topic, and there are many moving parts. It's easy to be discouraged by all the things that aren't working in the recycling ecosystem. The conclusion shouldn't be "Oh well, I'll just throw everything into the trash." Rather, the takeaway is that if you want to make a difference, then there's one surefire way to do that: *consume less*. As Kreigh Hampel, Burbank's recycling coordinator, recently told the *Los Angeles Times*, "Recycling is not going to undo the damage done by consumption."

Does Recycling *Increase* Waste?

Five hundred years ago, Martin Luther nailed his Ninety-five Theses to the door of the Wittenberg Castle church in present-day Germany. At least forty of his criticisms of the Catholic Church mentioned its practice of selling indulgences, decrees that fully or partially waived the punishment for a given sin. Luther's criticisms highlighted the notion that if someone could buy a "get out of jail free" card for sin, this would, in fact, cause people to sin more than they would have otherwise.

Those concerned about their carbon footprints today have available their own indulgences. They can buy carbon offsets—basically paying other people to plant trees to suck up the excess carbon generated by their behavior. Why not buy that gas-guzzling SUV as long as you have a few trees planted to offset it?

Back in 2007, Prince Charles, who has called climate change the "biggest threat to mankind," planned to fly to the United States with an entourage of twenty people to collect an award for . . . environmentalism. One wit at the time asked just how heavy the award was that he needed so many people to help him with it. The prince came under such intense criticism that he canceled a skiing trip to Switzerland to offset the carbon generated by his trip to the United States. Imagine the reaction you would get if, when confronted with the fact that your car might generate more pollution than average, you justified owning it by mentioning that you had canceled a ski trip to offset it.

In principle, there's nothing inherently wrong with offsets. However, they come under fire—and justly so—if they change human behavior for the worse and ultimately result in more waste and more net carbon in the atmosphere than if they hadn't been available in the first place. Because of this, there's long been a debate about whether offsets really accomplish what they purport to. Do they create *any* incremental change at all?

What about recycling? Is recycling itself just another manifestation of "indulgence buying"? John Tierney, *New York Times* science columnist, thinks so. In 1996 Tierney wrote an article for the *New York Times Magazine* entitled "Recycling Is Garbage," in which he argued that recycling simply wastes time, money, and natural resources. He asserted, "Americans have embraced recycling as a transcendental experience, an act of moral redemption. We're not just reusing our garbage; we're performing a rite of atonement for the sin of excess." Tierney followed the 1996 piece up with an op-ed in 2015 that carried much the same message. At the time the 1996 article ran, it received more hate mail than any article the magazine had ever run. The 2015 piece came under almost as much criticism.

What if Tierney is right, and recycling is a kind of atonement? Is there any harm in it? Maybe a little penance does us all a bit of good. Unfortunately, the analysis is not that simple.

Elsewhere in this book, we examine how different types of recycling can reduce the waste inherent in primary production of many goods, and often result in dramatically reduced energy consumption. Those chapters, however, only look at the first-order effects of recycling—they answer the question "All other things being equal, does recycling result in more or less waste?" In that context, the answer seems clear—recycling reduces waste.

But what if we ask a different question? What if all other things aren't equal?

Is it possible that recycling itself causes people's behavior to change? Can recycling actually result in more waste than would have existed otherwise?

Luckily, this is a pretty straightforward hypothesis to test, and two Boston University professors, Monic Sun and Remi Trudel, did just that in a series of experiments they reported in their paper "The Effect of Recycling Versus Trashing on Consumption."

In the first experiment, they gathered groups of people and told them they were participating in a study to compare the taste of four different fruit juices. They set up four self-serve containers, and next to each they placed a large stack of disposable cups. What they really wanted to know was whether subjects would use one cup multiple times to taste each juice, or whether they would get a new cup for each sample. In half the trials, they placed only a conventional trash can next to the table. For the other half, they put out a recycling bin. It turned out that the people who had a recycling bin next to them used nearly 30 percent more cups than the people who just had a trash can.

Then they did a different experiment. They got people to wrap a present and measured how much wrapping paper the subjects cut off the roll to use. It was the same setup: Half had a trash can next to the table, and half had a recycling bin. The people who had a recycling bin used 20 percent more paper.

In a third experiment, after a person completed a series of tasks, they were offered a free pen. The pens were in packages, and they were told to take as many pens as they wanted, and to dispose of the packaging before they left. Again, half saw a recycle bin, half saw a trash can. Those who had the recycling bin took 30 percent more pens.

These results are fascinating. Our responses are almost Pavlovian, except that when we see that blue bin, we don't salivate—we waste.

Where does that leave us? Do we conclude recycling is bad?

Not exactly, according to Trevor Zink, a professor at Loyola Marymount University whose views we revisit. Zink believes that consumption is our real problem. He says, "If people see that recycling is actually no more environmentally helpful than landfills, and we should feel appropriately bad about both, then perhaps they will say, 'Okay, maybe I'll consume less.'"

Zink's thesis, incidentally, goes way beyond mixed feelings about recycling. He maintains that fundamentally there's no such thing as "green" products and "not-green" products. Instead, there are only products that are bad for the environment and products that are slightly less bad for the environment. He argues that, in reality, "environmental damage happens at the point of purchase, in the production of the materials, not the disposal of the materials."

In the movie *War Games,* the character played by Matthew Broderick creates an experiment to show a sentient computer the futility of nuclear war. To do so, he instructs the AI to play against itself in games of tic-tac-toe over and over again. After millions of iterations, the computer concludes that tic-tac-toe is a "strange game" and "the only winning move is not to play." To Zink, the same is true of landfills, recycling, green products, and all the rest. To him, the only winning move is not to consume.

Of course, there are limits; for the moment, at least, life requires us to eat, to breathe, and, to a certain extent, to waste. While waste may be necessary, we should seek to limit it. Because when it comes to certain patterns of consumption, there can be no absolution.

Water, Water . . . Everywhere?

There's a quote that is variously attributed to one of three different sources—radio broadcaster Paul Harvey, the Farm Equipment Association of Minnesota and South Dakota, and Confucius—that goes, "Despite all our achievements, we owe our existence to a six-inch layer of topsoil and the fact that it rains."

Regardless of the provenance of this epigram, it's inarguably true. Water is central to life on earth. Land-dwelling animals never really parted from the sea—we're really just bags of seawater that acquired legs. In fact, the watery portion of our blood, the plasma, contains levels of salt and other ions shockingly similar to those of ocean water.

Viewed from space, our world looks like a water planet. Schoolchildren are taught that the earth is more than two-thirds water, and that's true if we consider only the globe's surface area, not its volume. But this notion is deceptive, since the water in question is little more than a thin veneer on the surface of a rocky planet. The earth's volume is 260 billion cubic miles, and the oceans make up only about a thousandth of that space.

Despite the fact that in recent years scientists have come to believe that massive amounts of water lurk underground (a volume perhaps one to three times the size of all the oceans), the earth as a whole is bone dry. Actually, it's even drier than bones. Bones are made up of 30 percent water, whereas the planet's water comprises only a tiny fraction of 1 percent of its volume.

Freshwater is even rarer. The oceans have around 300 million cubic miles of water, but the planet's liquid non-salt water is just 1 percent of that total. Even if we include all the ice on the planet, including those mile-thick glaciers covering Greenland and Antarctica, we find that a mere 3 percent of the earth's water is fresh.

Critically, it's that tiny bit of liquid freshwater that sustains each and every one of us. And so, in an almost subconscious acknowledgment of this reality, we're often warned not to waste it. But can water actually be wasted? The water we drink today is the same water that was around when the dinosaurs roamed the earth, and there's effectively the same amount of water on the earth as there was a hundred million years ago. To understand water waste we must consider something different from the destruction of water, because that doesn't really happen. However, from the perspective of its value to humans, water still can be wasted in that it can be contaminated or clean, salty or brackish or fresh. And, vitally, it can be "here" or "somewhere else."

To grasp all of the issues humans have with water, it's important to understand where the water we use comes from, and how much is available for our use.

Let's start at the beginning. Every day about 300 cubic miles of water evaporates from the oceans. That water becomes clouds, and those turn into rain. About 90 percent of that rain falls right back into the ocean, which seems like a whole lot of effort for nothing. But—and this is the important part—10 percent falls on land. If you spread that water over the 57 million square miles of land we have on this planet, over the course of a year you get about 11 inches of rain.

But the land also gets rain that doesn't originate in the oceans. Another 16 inches on average of the rain that hits the land originates from the earth's landmasses through a process called evapotranspiration. ("I'll take eighteen-letter words for $1,000, Alex.") Evapotranspiration refers to water that enters the atmosphere from soil and plants. In a process that's largely invisible to us, plants give off huge amounts of water. The leaves of one mature oak tree, for instance, give off 100 gallons of water a day, while an acre of corn gives off a staggering 3,000 gallons daily.

Water from evapotranspiration that later becomes rain bumps the average rainfall on the earth to a total of 27 inches. But overall, the water from evapotranspiration is a player in what's effectively a zero-sum game. Plants and soil give it off but then reabsorb it as well. Oak trees don't just magically emit 100 gallons a day from nowhere; rather, they also absorb 100 gallons a day from the land and air as well.

Thus, in our effort to understand water waste, we can effectively ignore evapotranspiration and concentrate largely on the extra rain that falls on land from the oceans. And like the water from evapotranspiration that's canceled out by what plants take back in, the 30 cubic miles of rain that fall on the land every day tries to make its way back to the oceans again. Water, as we all know, seeks its own level, which in the end is sea level. It always eventually ends up back home in the oceans.

Again, the oceans don't just give up a net 30 cubic miles of water a day to the land (the oceans would be empty by now if so). Rather, it's more of a loan that gets repaid. And this is the water we can siphon off, at least temporarily, to live our lives. This is the planet's *renewable freshwater*—an important concept here. It's the net water, after evapotranspiration, that falls on the land as well as the collective discharge from the world's aquifers, underground rock structures that contain or transmit groundwater.

To fully understand water waste, we need to introduce two

more concepts, those of water *stock* and water *supply*. Water stock is the total amount of freshwater in any given place at a given time; it consists of the water in lakes, rivers, swamps, and aquifers. Water supply is the amount of rain that falls (less evapotranspiration) plus the water that comes out of aquifers annually. Water stock is like a bank balance, the money in your savings account. Water supply is like your income, coming in over the course of a year; the income you don't use immediately can be added to your savings, or if your income doesn't completely cover your daily expenditures, you can take some out of your savings.

For instance, the Great Lakes of the United States are part of the water stock. They alone contain about 5,500 cubic miles of freshwater. That's nearly eight times the United States' renewable water supply, our water income. And just like with finances, it's always smart to spend less than your income, not dipping into your savings. Using up your water stock—in this example, draining the Great Lakes—isn't a good idea.

To see this dynamic in action, consider the Aral Sea. In the Soviet Union in the 1960s, the decision was made to take the rivers that ran into the Aral—then the world's fourth-largest inland sea—and redirect them into irrigation for cotton, a notoriously thirsty crop. This diversion resulted in the Aral's volume shrinking by more than 80 percent and the salinity of the remaining water increasing to a point where the lake is no longer a freshwater sea that supports freshwater fish and plants, but a saltwater one with a saltwater ecosystem. The area's vibrant fishing economy, which at one point employed sixty thousand people, dried up with the sea, and towns that had been fishing towns eventually found themselves 40 miles from the coast. To compound the disaster, salt left behind by the retreating Aral blew into good croplands, ruining them.

It won't come as a shock to learn that the water supply can vary dramatically depending on geography. There are billions of gallons of water supply for each person living in Greenland, but just a few hundred per person in Kuwait. Worldwide, about 25,000 cubic miles of rain falls on dry land each year. About 60 percent of that is given off by plants and absorbed by plants, leaving about 10,000 cubic miles as the planet's total renewable freshwater. An additional 3,000 cubic miles comes from aquifers, bringing the grand total of the water supply to about 13,000 cubic miles. Of that 13,000, Brazil has the most, at 2,000 cubic miles annually, or roughly 15 percent of the world's supply. Next is Russia, with 1,100.

The United States has a total annual renewable water supply of about 700 cubic miles. And every three days we withdraw 1 cubic mile of it through our consumption. That means that over the course of a year, Americans consume approximately one-sixth of their supply. That ratio of usage to supply puts the United States in the low-water-stress camp. In contrast, there are places— many, in fact—that suffer from extremely high water stress, which means they use virtually all of their water supply each year.

Globally, humans withdraw about 1,000 cubic miles of water per year against a renewable supply of 13,000 cubic miles, so the overall stress level on the planet is quite low. But like in the case of the man who drowned in a river whose average depth was 6 inches, this analysis is correct but misses the point entirely.

Global averages, while useful conceptually, don't shed much insight on the impact of water waste, because often the places where water happens to be are not where the people and farms are. To further complicate the picture, a good deal of the world's annual water supply comes in huge surges in times such as monsoon seasons and arguably shouldn't be included in the tally since there's no way to effectively harvest it. In addition, we choose to use much of the renewable supply to maintain navigable waterways; most places want rivers that run into the sea for transportation and recreation reasons. So while our planet-wide water stress level is low, it's not as low as it might seem.

What would a high stress level look like? Consider the river system of the United States. When the Righteous Brothers sang about lonely rivers flowing to the sea, they weren't talking about the Colorado River, which hasn't consistently made it to the ocean in decades, because we redirect all of its water into agriculture. The Colorado system is one in which 100 percent of water supply is used. Imagine if no rivers in the United States made it to the ocean, not even the mighty Mississippi. Roughly speaking, that would mean that the United States used all of its renewable water every year, and we would be considered a high-stress water supply country. Of course, water scarcity can vary quite a bit within a country; Las Vegas is under considerably more water stress than Seattle.

That's the backdrop against which we'll examine the waste associated with water in the next two chapters.

Use It or Lose It

When talking about water usage, there are two key concepts to keep in mind: water *withdrawals* and water *consumption*. A withdrawal is the temporary diversion of water from a stock source, like a lake or stream. Consumption is exactly what it sounds like: The water is permanently removed from the ground, a lake, or a stream.

What's the difference? If you're out fishing and you fill up a bucket to hold your fish, and then as you're leaving you dump that water back into the lake, you *withdrew* water, but you didn't consume any. You just borrowed it for a few minutes. But if you took that bucket of water home and watered your yard with it, you've consumed it.

Let's look at the United States as an example. In the United States, daily withdrawals are about 1,000 gallons per person. Consumption, on the other hand, is only about 300 gallons per person. To figure out the difference, let's look at why we withdraw water.

By far the largest two sources of water withdrawal are thermoelectric power and irrigation, roughly tied at 40 percent each.

Thermoelectric power is how we generate most electricity. We burn coal, oil, and natural gas to generate heat to turn water into steam, which turns turbines, which drive generators that produce electricity. Nuclear effectively works the same way—the nuclear reactions heat water into steam. All the machinery associated with this power generation needs a great deal of water to keep it cool. Once that water is hot it's of less value to the plant and is thus put back into the environment. A majority of plants in the United States use what's known as "once-through" cooling: Cool water is pulled from the environment, used for operations, and then discharged. Sometimes that discharge water is first cooled down in giant cooling towers that spray it in as mist. This huge need for water is why virtually all power plants are located near water sources. They withdraw as much water as agriculture, but they only consume about 3 percent of what they withdraw. Power plants *can* be made to withdraw much less water by cooling and reusing the same water. Then they would only have to withdraw what they lose to evaporation. But doing so would cause more waste in the form of less efficient energy generation from the same quantity of fuel.

Despite the relatively small amount of consumption, the diversion of water for electricity generation has certain problems. Pulling cold water from a river and replacing it with hot water kills aquatic animals. And, generally speaking, withdrawn water that returns to the environment isn't as pristine as when it came in.

The other massive use of withdrawn water is agriculture. In the United States, 400 gallons a day per capita are withdrawn. Unlike the case of thermoelectric power generation, consumption is quite high: roughly 65 percent.

You may be wondering how it is possible for irrigated water to be withdrawn but *not* consumed. If the water evaporates or if plants absorb it, it's considered consumed, because in both those cases the water leaves the local water system. But some amount of water will drip down through the soil back into the water table, and that fraction counts as withdrawn but not consumed. Because agriculture both withdraws so much and consumes a high percentage of the water that it withdraws, it accounts for 85 percent of all freshwater consumption in the United States.

That brings us to municipal water supplies. Of the 350 billion gallons a day that Americans withdraw from freshwater supplies, only about 40 billion—just over 10 percent—pass through a local water treatment plant. And of that volume, only about half is used in homes. The rest is devoted to commercial, industrial, and civic uses (such as lakes in parks). The roughly 20 billion gallons a day used by consumers works out to about 70 gallons per capita per day when we discount the roughly 15 percent of people who get their water from wells.

Of that 70 gallons, 20 are used outside, largely on lawns. This number is highly variable. An apartment dweller in Manhattan might not even water a petunia on their balcony, whereas a suburbanite in Santa Fe might maintain a golf course in their back forty. Of the remaining 50 gallons, people use about 20 percent each on the toilet, shower, washing machines, and various faucets. That leaves 10 gallons left over for everything else. Of that 10 gallons, half of it is lost—wasted—through leaks.

Of all the municipal water that's withdrawn, relatively little is consumed. Sewage is often treated locally and then reintroduced into local rivers and reservoirs. So you can flush your toilet a thousand times (whether it's a low-flow model or not) and you aren't really *consuming* much water, if any. While much of what is used on lawns is consumed, little of what you use indoors is. That's why when water supplies get low in cities they restrict outside car washing and lawn irrigation, which are consumers of water, instead of bathing, which only withdraws it.

Worldwide, water withdrawal and consumption numbers vary, as you might expect. In the United States, 1,000 gallons of freshwater are withdrawn daily per capita, but the world average is about 400. Of course, some places get by on very little. (Of note, countries' per capita water withdrawals don't correlate to income anywhere near as much as electrical use does, for less-developed nations often have ample water resources.)

This analysis skips over several other purposes for which water is withdrawn, including aquaculture, mining, and livestock cultivation, but these are minor in aggregate. In addition, there are many uses of water that don't count as either withdrawal *or* consumption. These include the creation of recreational lakes, water used for fish propagation, and navigation. Additionally, the 16 percent of the world's power that comes from the fifty thousand large dams spread around the world typically counts as having zero effect in terms of water usage, because water isn't actually diverted. (This may not be entirely fair. Consistent hydroelectric power usually involves giant reservoirs of water behind dams, which result in increased volumes of water lost due to evaporation.)

Since much municipal water is merely withdrawn but not consumed, theoretically a town could build a closed-loop system like the one on the International Space Station, where all wastewater is made pure again and run back through the same pipes. These "potable reuse" systems, sometimes pejoratively termed "toilet to tap," are in place in some areas, but their widespread adoption is limited by a few factors. The first is the cost. Generally, it's much cheaper to treat surface or aquifer water to make it potable than it is to treat sewage. That makes sense, at least for areas that have freshwater sources nearby.

The second thing limiting widespread adoption is the ick factor. Even though wastewater treated to be potable can be far cleaner than bottled water, a sizable minority of those surveyed say they would never drink it. Much of the issue is optics, since every glass of water you drink probably contains a molecule that has been peed out by some creature or another innumerable times, perhaps as recently as last week.

The fact remains that reservoirs, from which we pull lots of water, are full of living creatures and all that they produce. They contain dead and decomposing organic matter as well. In addition, treated water that's discharged is often put back into the same rivers and aquifers from which it was taken, only to be used by the next town downstream for its water supply.

Yet the ick factor is a real thing. There's enormous sensitivity around the issue of asking people to drink water, no matter how clean, that was recently sewage. When nineteen-year-old Dallas Swonger was alleged to have peed in a Portland, Oregon, reservoir—a charge he denied—Water Bureau administrator David Shaff decided to drain and discard all *38 million gallons* in the reservoir stating, "Do you want to be drinking someone's pee? . . . There's probably no regulation that says I have to be doing it but, again, who wants to be drinking pee?"

Situations like these are actually not that unusual. This is not a safety issue at all, and in this particular case nearby locals said they often saw dead animals floating in the reservoir. No doubt the birds flying above probably aren't "holding it in," either.

Beyond the cost and what we can refer to as PR challenges, there are additional issues to potable reuse. There is the question of how to dispose of "concentrates"—that is, the materials taken from the wastewater. There are legal and regulatory hurdles as well, the sorts initially put in place to protect the public, and a slew of political and technical issues.

The technical issues stem from the fact that the sorts of things found in wastewater in high concentrations are different from those found in reservoirs. It's easy to use filters to get bacteria and viruses out of water, but filtration can't get hormones, the molecules of which are two orders of magnitude smaller, out of the water. Wastewater has far more of these sorts of contaminants than reservoir water, and thus requires other processes such as electrodialysis or reverse osmosis, both of which use energy to pass non-potable water through a membrane while leaving contaminants behind, which are generally more specialized and expensive.

What would be less wasteful? Given that so much of the water withdrawn for irrigation is consumed, another conservation technique is to take wastewater and minimally treat it to use for irrigation. This system works well, and the treatment is much more affordable, but it does require a second set of pipes to handle the non-potable water.

Comparing the cost of potable and non-potable reuse is a bit tricky. Both cost less than desalination, and potable reuse is more expensive to bring online, but non-potable reuse requires its own piping. That's the big cost. A factory might easily reuse non-potable water in its toilets and for irrigating its grounds, but to install a second set of pipes in, say, New York City is clearly a different matter.

How about desalinating seawater as a source for freshwater? Doing so is generally the most expensive way to create usable freshwater. Sourcing clean water is a matter of cost, primarily in the form of energy use, which desalination requires a lot of.

Nevertheless, desalination is widely practiced around the world, with nearly twenty thousand plants in operation producing a cumulative quarter million gallons of freshwater every second. This particular solution makes the most sense for arid countries, which explains its popularity in North Africa, Australia, and the Middle East. Kuwait, for instance, gets 100 percent of its water from desalination. There are three problems with desalination, though. The first is that some ecological damage is done by sucking in vast quantities of ocean water. Second, the sludgy salt by-product of desalination has to be disposed of, and this is usually done by putting it back in the ocean, raising salinity (remember the Aral Sea?). Finally, and probably most significantly, there are the aforementioned power requirements.

By comparison, the energy requirements of treating freshwater and making it safe for human consumption are about one-tenth the cost of desalination. (This calculation doesn't count the energy needed to move water, which varies greatly by location.)

If we look to markets to figure out how best to allocate water and how to get the kinds of price signals that, in theory, can minimize waste, the picture can get more confusing. The retail price of water—your utility bill—in the United States is highly variable depending on where you happen to be and how much water you use, for reasons that aren't strictly correlated to supply and demand. Pricing per gallon used is often presented in tiers, with the price rising sharply as usage increases. This scheme is a de facto tax on lawns, as residential irrigation is the most common reason for high residential water usage.

In Dallas, for instance, your first few thousand gallons of water cost under $2 per 1,000 gallons, but once you get to a cumulative consumption level of 15,000 gallons, you pay over four times that amount for the same 1,000 gallons. As a general weighted average, Americans spend about $2 per 1,000 gallons of residential tap water. It's worthwhile to note that there are still a number of cities, some quite large, that don't meter water at all and instead charge residents a flat rate per month.

To achieve such reasonable prices requires significant subsidization on the part of municipalities. As a result, alternative sources of freshwater such as desalination often aren't economically viable—especially for people who use the minimum 15 gallons a day that is considered the amount needed for long-term survival. Additionally, water utilities are often constrained by statute to only charge enough to cover their costs, which include treatment and delivery. Doing so effectively values the commodity itself, water, at zero.

Of course, when "someone else" is paying, water subsidies often result in an overuse of water and prices so low that more ecologically sustainable sources can't compete with the status quo.

What effect did the COVID pandemic have on our water consumption? The short answer is that it didn't change all that much. Lawns still needed watering and toilets needed flushing. However, that's just the net effect. Zoom in a bit closer and you see something different. Water treatment facilities that primarily serve businesses saw demand plummet, while those that serve suburbia saw theirs skyrocket. It turns out we may have flushed the toilet the same number of times, but not in the same places. This shift in demand presents real problems for the water industry, whose systems are not accustomed to rapid long-term demand changes.

At this point we need to—excuse the pun—dive deeper into aquifers, for they are a big part of the story of water and waste. As mentioned earlier, only 1 percent of the water on the planet is liquid freshwater. That's what makes up all the lakes, rivers, swamps, aquifers, and all the rest. But here's a big surprise: Of the 1 percent that is liquid freshwater, *only 1 percent* of that is surface water, such as lakes and rivers; the other 99 percent is in aquifers. While all the freshwater lakes in the United States total about 20,000 cubic miles of water, swamps contain about 3,000, and rivers another 500, freshwater aquifers contain 2.5 *million* cubic miles.

Aquifers can be thought of as large underground lakes or incredibly slow-moving rivers. They discharge water naturally, often in valleys or into surface lakes. But they also experience another, artificial kind of discharge when humans pump water out of them. Sometimes, it should be noted, we pump water into aquifers during wet times to withdraw it later. But usually human pumping is only in one direction—out of the aquifer.

The water cycle, overall, is in balance if your time horizon is long enough. The amount of water that rains down on us is equal to what evaporates. Calculations of the water cycle often assume that balance, and they treat an aquifer's natural discharges and artificial discharges as the same thing. While this is true over geologic time, it isn't at all true on a year-to-year basis.

Renewable freshwater includes all discharges from aquifers. However, the majority of aquifers measure their recharge rate in centuries, not years. They have taken thousands of years to fill up, and the water in them today could be runoff from glaciers that melted at the end of the last ice age.

Early in this chapter, it was pointed out that societies should live off their water supply, not their water stock. So what of the aquifers? What are they? It depends. If you have an aquifer that quickly recharges and drains into a river, then that sure looks like water supply. It looks like a river. But if you have one that slowly fills up over eons, that looks more like the Great Lakes—like your water stock. In practice, however, all aquifer discharges are counted in water supply.

Because of this, aquifers act kind of like the credit cards of the water supply; we can cheat Mother Nature in the short term, and we do. In certain places, people use far more water than the amount of rain they get each year by dipping down into the water stock in the aquifers, year after year. And although we don't drain the Great Lakes, we certainly do pump from the aquifers. To-may-to, to-mah-to.

Consider the Ogallala Aquifer, one of the largest on the planet.

It's bigger than California and sits under eight Great Plains states. Since approximately the middle of the twentieth century, when humans figured out a way to economically adapt car engines into pumps to get the water out of the aquifer, it has helped feed the United States, and by extension, the world.

While it wouldn't make much sense to ship the water itself very far from the aquifer, it's extremely efficient to use the water where it is to grow crops and raise animals. In fact, one-sixth of the world's grain production is watered by this one aquifer. What's more, 30 percent of all the groundwater used for irrigation in the United States comes from the Ogallala. Four out of five people who live above it get their drinking water from it. We pump enormous amounts of water out of the aquifer, and in some places it has already run dry.

This borrowing, like a credit card balance, will eventually come due, and when it does, there will be a ripple effect through the whole region. Natural discharges will stop as well, drying out local lakes. One estimate predicts that by 2060, 200 miles of streams fed by the Ogallala will vanish. It will take over a thousand years to fully recharge this aquifer through rainfall.

It's hard to see how anyone could think of this aquifer as a renewable resource. As a result, unnecessary usage of water from the Ogallala Aquifer can very easily be thought of as waste, with serious consequences.

A minority of aquifers recharge quickly. Consider the Edwards Aquifer, which sits underneath eleven Texas counties. Every day 700 million gallons of water come out of it, half pumped by humans and half as the source of area rivers. Two million people, including parts of the populations of Austin and San Antonio, get their drinking water from it. Yet it remains full. During dry years its level falls, but during wet ones it recharges. The Edwards is a truly renewable resource. As a result, if one is to rank the impact of waste, leaving the water running while brushing your teeth in

Austin will ultimately have less of an impact than doing so in Wichita.

The challenge in understanding water consumption in general is that withdrawals from both the Ogallala and the Edwards are counted as coming from renewable freshwater, even though the rates of renewability vary considerably. Across the entire United States, of the 350 billion gallons of freshwater withdrawn a day, a quarter of it comes from groundwater. No one is entirely sure just how renewable that water is in aggregate. In the United States, the largest user of groundwater is California, which pumps 17 billion gallons a day from aquifers; according to Stanford University, that accounts for two-thirds of the state's water needs during dry times. These withdrawals allow places like California's Central Valley to grow food in more abundance than it would be able to do without the groundwater.

It's easy to cast stones and say, "Maybe they shouldn't be farming in the Central Valley if it takes a gallon of water to grow one almond, and we have to pump part of that gallon from an aquifer." That analysis is incomplete, however. It's true that growing almonds in the Central Valley requires quite a bit of water, but there are many other inputs in growing crops. Take soil. The Central Valley contains the single largest expanse of Class 1 soil—the Cadillac of dirt—on the entire planet. It's a nearly perfect growing environment, exempt from wild temperature swings and blessed with ample sunlight year-round.

Water policy in the United States has always been (at least theoretically) about taking water from where it's of little use and moving it to where it's of great use, and the Central Valley is an example of that practice—even if it involves rerouting water over the Continental Divide for irrigation, which it does. And problems emerge if the water use isn't sustainable, which it isn't.

Worldwide, we've seen aquifers go completely dry after a single generation of pumping. NASA uses satellites to measure localized gravity levels to estimate the volume of water in the world's aquifers. They believe that of the planet's thirty-seven largest aquifers, twenty-one are running dry.

One visible result is sinking cities. Consider Mexico City. The greater Mexico City area has a population of over twenty million. It gets a pretty respectable rainfall of 25 inches a year; but 85 percent of that rain comes in five wet months, which leads to flooding and an inability to capture all the water. The rest of the time the Mexican capital faces shortages so severe that 70 percent of the population only has running water twelve hours a day. A fifth of the population relies solely on water delivered by trucks; anecdotally, some families spend a quarter of their income buying water. At an elevation of a mile and a half above sea level, water has to be pumped up to the city at great expense, and the water infrastructure is so old and decrepit that 40 percent of all water is lost to leaks and theft. To supply the city's water needs, residents pump up water from an aquifer at a rate twice as fast as it can be replenished. The result is that the city sinks by more than a foot a year as the aquifer empties out.

Beijing, another city of twenty million, sits atop an impressively large aquifer that's also being drained dry. Beijing sinks 4 inches a year, which, among other things, could cause train derailments, since the sinking is far from uniform. China is trying to mitigate this problem through one of the biggest, most ambitious engineering efforts of all time, a $60 billion effort to pipe 1 million gallons of water *every three seconds* from the wet south to the dry north. To put this endeavor in perspective, one of the world's largest oil pipelines, the Druzhba in Russia, delivers a mere 1,500 gallons of oil in the same amount of time.

This redirecting of water to the population and farming regions has been occurring on an ad hoc basis for decades in China, and by one estimate (disputed by the Chinese government) has resulted in the disappearance of half of China's rivers.

These two sinking cities are not isolated examples. Every two years Jakarta sinks a foot due to overpumping of groundwater. Houston sinks a foot every six years for the same reason. And the lower coastal cities go, the more likely it is that hurricanes and other storms will damage them. In the last century, the entire San Joaquin Valley in California has sunk 30 feet—and is still going down.

A Deep Dive into Water Waste

With a solid grounding on how the relationship between humans and water works, let's bring it all home and explore exactly why the world has a problem with water and waste.

As previously noted, water itself isn't scarce. Even freshwater isn't a limited resource, if you think about it. A virtually infinite amount of water can theoretically be piped anywhere on the planet. With desalination, it can be made fit for consumption. The challenge is the energy needed to do so, which equates to cost.

From one perspective, there's no problem at all. Unlike fossil fuels, we have enough raw material to meet all of the demand we're ever likely to have for water. All it takes is an expenditure of electricity to clean it. Further, cities could effectively reuse the same water over and over again by applying technology and energy. So, is there *really* a problem with wasted water?

Yes, a big one.

You can see it in the numbers. Half of the people in the world experience extreme water scarcity at least one month a year. One in ten people on the planet doesn't even have access to basic non-potable water services for washing and agriculture. Three in ten lack access to safely managed clean drinking water. The UN estimates that over the next decade, 700 million people might be displaced by water scarcity. Poverty is the root cause of much of this phenomenon. But the economics are compounded by five additional unique problems with water.

First, water is in the "wrong" places. Water scarcity is entirely a matter of location. Urban areas and farm regions aren't necessarily located near large water sources. Think of Los Angeles, whose annual rainfall, about 11 inches, is just an inch above the amount for areas that are considered deserts. Or Las Vegas, for that matter, which gets but 4 inches a year. And it's not just Las Vegas. In an episode of the animated TV show *King of the Hill*, Peggy Hill and her son Bobby, dyed-in-the-wool Texans, make their way to Phoenix, Arizona, and have this exchange:

> BOBBY: "111 degrees? Phoenix can't really be that hot, can it? Oh my god, it's like standing on the sun!"
> PEGGY: "This city should not exist—it is a monument to man's arrogance."

When we reconsider the story of the drained Portland reservoir mentioned earlier, perhaps we shouldn't mourn that lost water. Water Bureau administrator David Shaff, who made the decision to drain the reservoir, had this to say: "It's easy to replace those 38 million gallons of water. We're not in the arid Southwest; we're not in drought-stricken parts of Texas or Oklahoma." His point is well-taken. With its 43 inches of annual rainfall, Portland is hardly the desert planet Arrakis from Frank Herbert's *Dune*. When you have massive rivers coursing through your sparsely populated part of the country, what's the harm in emptying 38 million gallons into one of them and then drawing another 38 million gallons out? Shaff may have overreacted, but it wasn't a tragedy.

Many population centers, however, are in areas that don't have ample renewable freshwater sources. Cities whose main water supplies are aquifers that take a long time to refill are ticking time bombs. These cities also require food, so there are often nearby agricultural areas that utilize non-renewable groundwater for irrigation in areas not otherwise suited for cultivation. Ironically, access to aquifers makes water problems worse in the long term. When there is plenty of water, people move into an area and start farming and ranching. Population grows. But then one day the aquifer runs dry.

Second, water infrastructure is either bad or absent in many parts of the world. The UN reports that 40 percent of people in the world don't have a basic handwashing station at home—that is, running water and soap. A quarter of the world's *healthcare facilities* lack basic drinking water services. The impact of this scarcity is almost beyond reckoning. According to the Water Project, if everyone had clean water, the number of people in hospitals would fall by half.

Making the problem worse, what infrastructure exists is often leaky, and those leaks are hard to track down and repair. In South Africa, over a third of all the country's scarce water is lost to leaks. In Mumbai, it's more than that. These places are not extreme outliers.

Third, the economics of water ownership often leads to less-than-optimal outcomes. Economist Garrett Hardin developed the concept of "the tragedy of the commons," in which people collectively overconsume a scarce resource because they're individually incentivized to maximize their own consumption of it. Think about people who fish for shrimp. It's in none of their interests for shrimp as a whole to be overfished, but it's in everyone's individual interest to personally overfish. One solution is for government to step in and strictly enforce a limit on daily catches. Problem ameliorated.

The tragedy of the commons is a big force in water allocation. As noted earlier, water's economic value is clearly not zero, but consumers are often billed for water as if it is, which prompts overuse. Surface water irrigation rights were often doled out a century ago, and the landowners who have them can and do withdraw the water they're entitled to even if there may be a higher-value use down the river. Why wouldn't they, if there's no reasonable mechanism for them to be rewarded for not doing so?

The situation is compounded with regard to pumping ground-water from aquifers. If no one owns the water, no one has any disincentive not to use as much as they want; after all, they reason that their neighbors probably are doing just that. Economics has few lessons clearer than this one. When the price of a normal good falls, people consume more. When it's free, people have no reason not to waste it.

The idea that free water is overconsumed is not a new observation. We know many stories about it from antiquity. People have been using aquifers for irrigation and drinking water for thousands of years with an amazing degree of sophistication. None did it better than the ancient Romans, whose mastery of practical engineering can be seen even today at a thousand archaeological sites. Their aqueducts moved massive amounts of water great distances—as much as a hundred miles—and the construction was so precise that the downhill slope of an aqueduct might be only 1 inch per mile of length. Seventeen hundred years ago, eleven aqueducts provided Rome with about 3,000 gallons of freshwater *per second*. Assuming about a million people in the city at the time, that works out to about 250 gallons per person per day, an amount on par with many countries today.

In an ironic turn of history, due to a water shortage, present-day Rome recently had to shut off the public drinking fountains that flowed freely in Cicero's day. The present-day Roman water system is also plagued with problems, including perhaps 50 percent leakage, and water pressure has been so low that some top-floor apartments have had to haul water up in buckets. In Italy as a whole, a quarter of all water pipes are more than half a century old.

If free water leads to overconsumption, is there a way to charge farmers for the water they pull from aquifers to irrigate their crops? Wouldn't doing so result in more efficient usage?

In theory, yes. But doing so is a hard sell to farmers, many of whom barely scrape by as it is. To be told that they will now have to pay for water they pump from wells on their own land is a bitter pill to swallow. However, this strategy is being tried in various places around the world. One example is in the San Luis Valley of southern Colorado. Due to a shrinking aquifer, when water meters were placed on wells, farmers began paying about 25¢ per 1,000 gallons of usage. The price has since doubled. Funds raised are then used to pay other farmers to leave their land fallow. For agreeing not to pump water into a 120-acre patch of land that previously had been irrigated, a farmer can receive $200 an acre per year. The local water authority is trying to implement this scheme across 40,000 acres, but adoption has been slow so far and they're only a quarter of the way there. The hope is that with 40,000 acres taken offline, the aquifer will stabilize. It's too early to tell if this particular program will work.

The fourth circumstance that leads to water problems is the ease with which "virtual water" can be exported. To understand what virtual water is, consider this story about Saudi Arabia told by reporter Nathan Halverson. That desert country sits atop an aquifer that was recently the size of Lake Erie. It took more than ten thousand years to fill up, and it has provided water to many cities as well as to oases scattered across the country, some so ancient they are mentioned in the Bible. For thousands of years, the aquifer made life—and civilization—possible in this area. But that all changed when the Saudis decided to get into the wheat business. They began pumping 150,000 gallons per second from the aquifer. Overnight, Saudi Arabia became the sixth-largest exporter of wheat on the planet. That party ran for twenty-five years. Then the water was largely gone. The wells dried up, as did the wheat crops. Saudi Arabia now desalinates half the water it uses, and for most of the rest continues to pump from ever-deeper wells from the 20 percent of the aquifer that remains. One way Saudi companies have responded is by purchasing up large tracts of land with water rights around the world and growing alfalfa—a water-intensive crop no longer allowed to be grown in Saudi Arabia—for export back to Saudi Arabia to feed cattle. They are essentially exporting water from places like central California that themselves have precious little of it. That's virtual water. In an NPR interview, Halverson talks about a purchase by Saudi food company Almarai of 15 square miles in the Arizona desert. He remarks, "They got about 15 water wells when they purchased the property. Now, each one of those wells can pump about 1.5 billion gallons of water. It's an incredible amount of water they're going to be drawing up from that aquifer underground."

But the Saudi case is just an example. Virtual water is trafficked everywhere. For instance, it takes a great deal of water to produce beef or corn. And when you export those products, you're exporting virtual water. There's a huge flow of virtual water from North Africa to Russia in the form of citrus crops, which grow in abundance in Morocco and Egypt.

It's by this mechanism that water can leave places that desper-

ately need it due purely to economic forces. Just like the phenomenon by which 80 percent of all hungry people live in
countries that export food, many thirsty people live in places
that export vast quantities of virtual water. If the world market
will pay more for it than the local people can afford to pay, the
water leaves.

Then there's the fifth problem, the reliance of modern agriculture on huge amounts of irrigation. Is there a solution to this?
Could we grow food with just rainwater, without irrigation, or at
least without using groundwater? Let's break this question down
into parts.

First, let's examine the question of whether we can grow our
food *the way we currently do* using rainfall alone. Almost certainly, the answer is no.

The U.S. Geological Survey states it this way: "Large-scale
farming could not provide food for the world's large populations
without the irrigation of crop fields by water gotten from rivers,
lakes, reservoirs, and wells. Without irrigation, crops could never
be grown in the deserts of California, Israel, or my tomato patch."
And that's in the United States. Bangladesh and Pakistan, to take
two examples, irrigate more than half their pasture and crop-
land. India irrigates a third of its.

The challenge is that irrigation at the margins is sometimes
what it takes to make a crop successful. Corn, for instance, needs
22 inches of rain to achieve a good crop. A low-yield crop can be
had with 15 inches. If rainfall is bad in a year, irrigation is the difference between no crop and some crop. Exacerbating the problem, irrigation encourages the production of crops with high
water needs. Beans need a mere 12 inches of rain, as do peas.
Wheat and sorghum need 20 inches. Cotton and rice need 40.
With no water for irrigation, farmers might still farm just fine,
but by growing completely different crops. While we cannot

know just what agriculture would look like without groundwater irrigation, we can confidently say that our diets would be different.

But let's reframe the question again. The key phrase in how we asked the question was around the way we currently do agriculture. Could we change the whole game and grow our food with a tiny fraction of the water we currently use? Yes. As you will see in the chapter on wasted food, the second-largest food exporter in terms of value is the Netherlands, a country not much bigger than Maryland. The Dutch, with their high-intensity, multilayer greenhouses, use but 15 pounds of water to grow a pound of tomatoes, whereas in the United States and most of the rest of the world it takes 80 pounds or more. The Dutch achieve this water efficiency using a cornucopia of technology, from carefully controlling the composition of the air in their greenhouses to using robots and drones. Every plant is like a beloved child, carefully nurtured every day.

We can also use other technological tools. We can genetically modify our crops to require less water. We can rethink how we irrigate and fertilize and all the rest. We can employ precision farming and nanotechnology. And we can alter the soil's microbiome. There are literally thousands of patents issued in agriculture every year, and with good reason, given that in the United States alone the output of our farms is valued at over $130 billion annually.

There really is a way to feed the growing population of the planet using less, including much less water. But it's hard to see how a country with "free water" to use in irrigation would make the transition to high-tech indoor farming like the Dutch do. And of course, all of this is only theoretically possible in the parts of the world that have the resources to make the kinds of capital investments needed to create the high-tech "farm of tomorrow."

Those are the five big problems—the water is in the wrong place, infrastructure is bad or missing, the overconsumption of "free" water, the ease with which virtual water can be exported, and the reliance of modern agriculture on irrigation. Given those problems, what can we do as individuals to help?

While one person's ability to make a huge impact is limited, there are three good possibilities for how an individual might help. The first is conservation. The kinds of practices that conservationists might think are hugely impactful are, while not completely ineffective, perhaps less useful than one might think. Installing low-flow toilets or taking shorter showers has limited impact, given that residential water use is such a small part of overall withdrawals and an even smaller part of consumption. Sure, do all the commonsense things. If you live in a water-stressed area, don't landscape your yard with thirsty St. Augustine grass and don't wash your car with a hose on the street.

But herein lie the limits of residential water conservation. When it comes to items like plastic bags or water bottles, consumers are collectively the direct users of those items. If you're concerned about them, stop using as many. If everyone does so, then, well, problem solved. But water use is completely different. Given that unfathomably large amounts of water are still used to irrigate crops that would otherwise not grow in an area, it doesn't matter much if you leave the faucet running while you brush your teeth or whether restaurants bring you water only upon request. Maybe growing water-intensive crops in arid climates isn't a great idea, but unless you're a farmer doing that, you don't have much say.

Perhaps you're thinking, "Okay, I will just stop buying those water-intensive products." That's a second possibility—changing your consumer behavior away from water-intensive items. Vote with your wallet, as it were. But this tactic is problematic as well.

You've probably seen all those statistics that purport to tell you how much water is used to make a thing. We are told, for instance, that it takes 1,000 pounds of water to grow a pound of wheat, 4,000 pounds to make a pound of rice or a pound of sugarcane, 20,000 pounds to make a pound of cotton, 5,000 pounds to make a steak, and 200 pounds to produce a cup of coffee or a glass of wine. Sometimes the numbers are reduced down to gallons per calorie of food. A calorie of beef, for instance, takes 2 gallons, while a calorie of chicken takes less than 1 gallon. Potatoes only need 2 cups of water to make a calorie, and the same goes for cereal grains.

These numbers are true. And they're put out by organizations in good faith. Much time and academic rigor are spent figuring this all out. But by their very nature these calculations are too simplistic to be especially meaningful, for they count different sources of water equally. A cow raised in water-rich Oregon that feeds on unirrigated grassland has an entirely different impact than a Saudi cow raised on alfalfa that was imported from the Arizona desert—or a Hawaiian cow that was raised on the islands but flown to the mainland on a 747 (which actually happens). The geography of where the water was sourced from is hugely important. But even if you know that, it doesn't tell you much because there's a huge variation in impact between rainwater (which is counted in those "number of gallons needed" numbers), diverted surface water, non-renewable aquifer water, and renewable aquifer water.

Try to imagine a labeling system for items that attempted to capture all that nuance. It would be nearly impossible to say anything meaningful even about something made of a single ingredient, like a cotton T-shirt. Just imagine tallying it up for a pizza. And even if there was such a system, the metrics of any given product would be constantly changing and impossible to verify.

So, sadly, we must put aside our hope for a simple apples-to-apples comparison of water usage. What we really care about is not how much water went into a thing but how much that water usage impacted the world, and that's a hard thing to know. Today, if you want to make sure the cotton in your clothing was grown entirely using rainwater, good luck figuring that out. That would be as difficult as making sure that the gasoline you put in your car wasn't made from oil extracted using fracking. The world just isn't set up to measure those sorts of things.

Okay, how about water offsets? You can buy carbon offsets, right? Why not buy water ones? This too is problematic. If you emit carbon dioxide in Akron and offset it by having trees planted in Ankara, good for you. We all breathe the same air, so it doesn't really matter where your carbon offsets occur. But if you live in the desert and decide to waste 38 million gallons of water for the spectacle of it and then to assuage your conscience decide to buy water offsets in Portland, don't expect any medals. They have so much freshwater up there that they'll throw away 38 million gallons if someone pees in it. Water scarcity is incredibly place-dependent and time-dependent.

Is that really the way it is? We have a worldwide water problem and there's nothing we can individually do about it? After such a long journey, it seems like an unsatisfying place to leave things.

Maybe the situation isn't quite that bad. There are three things that can have a real impact. You can change your consumption habits broadly away from things that in general require more water, such as eating beef or buying new cotton clothing. Second, widespread knowledge of how the whole water ecosystem works is lacking. If more people understood the problems with pumping from non-renewable aquifers or just how leaky our water infrastructure is, then things can change. And finally, many issues around water have policy implications, such as water ownership rights. Making sure elected officials are conversant in these issues would go a long way toward finding solutions.

But that's mostly it. Maybe that's why there's so much focus on low-flow toilets and the rest. Maybe they don't do much in the grand scheme of things, but it's something people can do that helps . . . a little.

PART 2

Waste in Our Business

Oh, You Have the Old Version

An early draft of this book had a line that, in retrospect, turned out to be naively hopeful: "No one is in favor of waste." But in the writing of this book, we've been thoroughly disabused of that notion. Waste has countless advocates.

After all, with more waste, more money winds up in the pockets of many people and businesses. The more we throw out canned goods because the "best by" date has passed, the better it is for purveyors of those goods, even though that date is not set by any objective authority and has nothing to do with food safety. "Better safe than sorry," many people reason, not realizing that their misplaced sense of caution often causes them economic harm.

This phenomenon is pervasive. College textbook publishers put out new editions of their books that differ only slightly from prior editions, thereby rendering used copies of the text obsolete. When showering, do you really need to rinse and repeat? Shampoo makers would have you believe so.

In fact, very rarely do the manufacturers of *anything* shed a tear when customers become dissatisfied with last year's version of a product and buy a newer one. The comic strip *Bloom County* once ran a storyline where the local computer hacker, Oliver Wendell Jones, saw a TV commercial touting the new version of his computer, which was exactly the same as the one he had except that the new one featured one update: "tint control." The punch line panel showed the old computer in the trash, with Jones remarking, "Hackers, as a rule, do not handle obsolescence well."

No, they don't. And neither do a great many others. Hemlines rise and fall. Men's ties get wide and then get thin. Portable phones get tiny and then enormous. If breathlessly chasing after each new fashion is something you want to avoid, then with enough patience you can wait the cycle out and all that was old will become new again. In the meantime, you must bear the shame of wearing ties whose widths are out of style.

What do these examples—from dates on cans to tint control to tie width—all have in common? They are all about obsolescence, the process by which things become outdated or unusable. But there's a particular type of obsolescence that deserves our attention: planned obsolescence, or the deliberate design of products to either become unusable, impaired, or out of fashion at some point before their useful life is over. Is planned obsolescence waste? At first glance it would seem to be about as wasteful as anything can be.

Planned obsolescence comes in two main varieties. Engineered obsolescence describes when products are purposely designed and built *to fail* before they otherwise should. Style obsolescence, on the other hand, describes when products are designed *to go out of fashion* on a regular basis. Let's begin our analysis with the first one, which is seemingly the most pernicious.

In the movie *Saving Private Ryan,* Tom Hanks's character, Captain Miller, reflects on the cost he and his men will pay to bring Private Ryan home, and remarks, "He better be worth it. He better go home and cure a disease, or invent a longer-lasting lightbulb."

Ah, the lightbulb: a product so beloved it has become the universal symbol of a good idea. It was almost as perfect as a product could be, but it had one glaring flaw—it burned out.

The lightbulb features prominently in nearly any discussion of planned obsolescence because of the actions of a small group of lightbulb manufacturers that came to be known as "the Phoebus Cartel."

For those lacking encyclopedic knowledge of the history of electric illumination, in 1924 the world's lightbulb makers, worried about slumping sales, gathered in Switzerland and agreed to design their products so that each bulb would last for only 1,000 hours, which was less than many of the bulbs already on the market. Not only did they set a target, but they imposed a system of fines for companies that made lightbulbs designed to last longer. As a consequence, engineers labored day and night not to make bulbs last longer but to ensure that they burned out more quickly. The actions of the Phoebus Cartel are often touted as some of the earliest and best-documented attempts to increase profits through collusion and planned obsolescence.

But the story isn't quite as cut-and-dried as it's usually depicted. Good reasons exist to limit the life of lightbulbs. Even today, incandescent lightbulbs are already massively inefficient— they turn just 3 percent of the electricity they consume into light, with the other 97 percent being converted into heat. To give a bulb a longer life requires either increasing that amount of inefficiency by using a thicker filament or lowering the brightness of the bulb, requiring more bulbs for the same amount of light. It's reasonable for lightbulb makers to design their products in such a manner as to standardize the balance of efficiency and longevity, particularly when the cost of the product itself is dwarfed by the cost of the energy needed to power it. Agreed-upon standards can make sense. And while today's government often gets involved in setting such standards, it wasn't common practice in 1924.

Of course, such a charitable reading of the phenomenon swings the pendulum too far in the other direction, replacing the lightbulb makers' devilish horns with saintly halos. There is no doubt that the cartel had its eyes squarely on the bottom line when they met in Switzerland. The efficiency argument had the benefit of being both true *and* a good cover story. But human motives are seldom entirely selfless or selfish.

Nevertheless, the idea that we live in a world full of products designed specifically to be shorter-lived than they could be, at the same cost, is a hard one to shake. Planned obsolescence has become a rare example of a conspiracy theory that exists in the mainstream, rather than on the fringes. Who hasn't wondered, when something breaks right after the warranty expires, whether the manufacturer engineered that outcome?

The idea that products are built to fail in order to force people to buy new ones seems plausible—the sort of scheme that would have been concocted by a portly nineteenth-century captain of industry, or perhaps Scrooge McDuck. Even dystopian science fiction explores the idea. In 1982's *Blade Runner,* the Tyrell Corporation built replicants with a four-year life span, ostensibly for public safety reasons, but with the added benefit of being able to sell new replicants on a regular basis.

Indeed, an oft-told story about Henry Ford alleges that the industrial titan sent teams of engineers to junkyards to look at discarded Model Ts, with the goal of finding parts on them that never failed. When they returned, Ford's teams reported that they had found one: a part known as the kingpin. To the engineers' surprise, Ford ordered them to make a lower-quality, and thus cheaper, kingpin, concluding that the one in the Model T was overengineered. It's a good story—but it's not true.

True stories of engineered obsolescence are difficult to identify, which is to be expected since it's the sort of thing a company would keep secret. In theory, as the example of lightbulbs made clear, a free market system has safeguards against overtly engineered obsolescence. A company that intentionally releases a product with flaws will be displaced by another that comes along with a better version to win away customers—the proverbial better mousetrap.

Still, the waste that goes into obsolescence does seem to be deliberately engineered into our everyday lives. We see it in printer cartridges that could theoretically be refilled, but which contain chips to disable them when not using consumables made by the manufacturer. A trend has emerged to make products whose batteries can't be changed out, or with no user-serviceable parts, or that are designed to require special tools or software to work on. Software companies themselves make the costs of switching to their competitors virtually insurmountable, not through deliberate engineering, but simply by not supporting the ability to export information into a standard format. Many technology companies deliberately design their products with proprietary features that lack value simply to make them less interchangeable with their competition.

How can companies get away with this? Why aren't their competitors eager to make better products and knock them off their perch? The answer is simple: There are many companies that hold such commanding leads in their industries that they can, for a time, exploit that lead to make purchasing their products a practical requirement. The market leader will employ a thousand contrivances to protect its dominant position. What dry cleaner wouldn't want to be the only one in town? Market leaders can indulge in what economists call monopolistic pricing, to the detriment of consumers, and build products in a way that favors their own bottom line over their users'.

And yet, products still get better. Entrenched companies go out of business. How? While it's the sport of market leaders to protect their leads, it's the sport of the upstart to knock off Goliath. Netflix streamed movies, and Blockbuster vanished. Manufacturers made cameras that didn't use film, and Kodak and Polaroid filed for bankruptcy.

So that's engineered obsolescence. What about the second kind of planned obsolescence, style obsolescence? Tail fins on cars, while a distant phenomenon, are probably well known enough to be a good example. This year's model has bigger, or smaller, or longer fins, and if you don't have them, you're out of style.

Engineered obsolescence has no real public defenders beyond the companies themselves. What might surprise you is that stylistic obsolescence has many defenders. Let's hear them.

In the 1930s, an American man named B. Earl Puckett did what we would call today a "roll-up" of department stores into a chain called Allied Stores Corporation. In the middle of 1950, at the Astor Hotel in New York, he addressed an audience of about five hundred fashion merchandisers who provided his stores with their merchandise. He explained what he wanted from their products, in a way that was appalling to our modern sensibilities: "We in the soft line business have the responsibility to accelerate obsolescence. It is our job to make women unhappy with what they have in the way of apparel."

He continued, "We must make these women so unhappy that their husbands can find no happiness or peace in their excessive savings. We have not made our proper contribution to the total economy until we have achieved this result."

Think about that. Puckett asserted that stylistic obsolescence contributes to the economy.

A more eloquent defender of stylistic obsolescence was J. Gordon Lippincott, a giant in the design world. Lippincott designed the Campbell's Soup can and made the Coca-Cola logo iconic. He designed the inside of the Nautilus nuclear submarine as well as the Tucker automobile. He put the god Mercury on FTD's logo, the spoon on the Betty Crocker logo, and the *G* on General Mills products. He was also a nuanced thinker about corporate identity, and in 1947 he wrote a still highly readable book on the topic called *Design for Business*. In it, he lays out his argument for stylistic obsolescence. It begins with an observation:

> Our custom of trading in our automobiles every year, of having a new refrigerator, vacuum cleaner or electric iron every three or four years is economically sound. Our willingness to part with something before it is completely worn out is a phenomenon noticeable in no other society in history. It is truly an American habit, and it is soundly based on our economy of abundance. It must be further nurtured even though it is contrary to one of the oldest inbred laws of humanity—the law of thrift—of providing for the unknown and often-feared day of scarcity.
>
> When an automobile becomes style-obsolete it moves down the line to the second-hand car buyer and continues a useful life until it finally hits the graveyard and becomes scrap metal for re-use in industry. I insist it does not matter whether it becomes junk in the hands of the second owner or the fifth or sixth owner down the line; the important point is that, if the original owner used this car for its full life of 15 years there would be no car for that fifth or sixth owner.

Lippincott makes a good point here, one that's still relevant today (as we discuss in the chapter on cellphone recycling). If you buy the best new smartphone and then, two years later, buy the better and newer one, you probably don't throw the old one away. Perhaps it goes to one of your kids or you list it on eBay. Perhaps it's sold back to the manufacturer and takes on a new life in the secondary market. In this case, Lippincott is right. Stylistic obsolescence can actually serve a useful purpose: It creates a secondary market, to the benefit of a group of users who are either unable or unwilling to buy a new phone. If we all kept and used our phones until they were useless pieces of junk, the phone market would undoubtedly be smaller, and many people would likely be kept out of the market altogether. If the only cars available were new cars, many more people would be riding bicycles.

But let's not lean on this one example too hard. As you'll see elsewhere in this book, 80 percent of donated clothes end up in landfills. Many, if not most, electronics in dumps were functional when they were discarded. So we need to bifurcate our analysis of the waste involved in stylistic obsolescence into cases where the product gets passed to a secondary market and those where it gets thrown away. If an out-of-fashion product runs out its useful life, then what's the harm? If it doesn't, that looks like waste.

Lippincott has a second justification of stylistic obsolescence that he offers in *Design for Business*. He asserts that the demand for novelty, whether organic or engineered, "stimulates continued effort on the part of the manufacturer to give the consumer something better, and as such is a means of raising our standard of living."

Smartphone vendors, as an example, have annual models, just like car manufacturers. And their goal has to be to improve the phone enough that owners of last year's model see a need to upgrade. In the early days of smartphones, these updates were relatively easy to engineer, but they're becoming an increasingly tall order. No doubt the manufacturers of these phones work ever harder to come up with new features, and this process drives innovation.

In fact, this cycle can actually be thought of as a third kind of obsolescence: *innovation obsolescence,* or the process by which an item is made obsolescent by an objectively superior version of itself. A cheaper and more effective vaccine, for example, might well render obsolete all the current inventory, but it's hard to see this practice as wasteful (even though the existing vaccine is still functional).

It should be noted that Lippincott ended up recanting many of his views just a few years later. He wrote an article railing against the annual model changes of cars and other durable goods. His arguments against constantly changing products were numerous. Style changes become merely change for change's sake and don't improve products; in striving for "improvement," products that begin with excellent design are actually made worse; constantly retooling manufacturing plants increases prices and takes energy away from improving product quality; relentless change and new models make for buggy products; repairs become more difficult as models proliferate; the pressure for novelty leads to ever more complicated products that are harder for consumers to use; and finally, consumers grow jaded in a world where *every* product is touted as new and improved. (Lippincott didn't point it out, but nothing can be both new *and* improved.)

One final defense of planned obsolescence deserves attention: the notion that planned obsolescence powers an economy. Proponents of this theory suggest that constantly having to replace things leads to growth and employment. If nothing ever wore out, no one would buy anything other than consumables. This proposition is often framed as a scathing critique of capitalism, asserting that, at its core, capitalism requires ever more raw materials to power the economy, which fuels the exploitation of nature.

But whatever one thinks about planned obsolescence or capitalism, this particular criticism isn't valid. It's the embodiment of a famous fallacy in economics called the broken-window fallacy. In it, a boy throws a rock through a window, breaking it. The townspeople, upon reflection, decide that the boy has done a good thing. Now the glassmaker has work where business was slow before. He in turn hires a laborer to install the window. The laborer needs transportation to get to the job site, so he takes a cab. The rock through the window turns out to be an economic boon—or so the fallacy asserts.

But as Frédéric Bastiat pointed out in his essay "That Which Is Seen, and That Which Is Not Seen," the money used to pay the glassmaker would have been used for something else—perhaps window shades—stimulating the economy to the same or an even greater degree. The only difference is that with the broken window, the homeowner doesn't get ahead; she is simply restored to her prior status, whereas another use of the funds would have been a net gain.

The reasoning extends to planned obsolescence. A window designed to randomly break isn't good for the economy, for the same reason. Capitalism no more requires windows that break and need replacing than it does anvils or caskets that do likewise. If someone does eventually make and sell a window that can't be broken, it's a great thing from a capitalist standpoint because window-repair funds will be freed up to create new jobs elsewhere.

The idea behind the broken-window fallacy is so seductive that there have even been proposals to implement it through legislation. In 1932, a man named Bernard London wrote an essay called "Ending the Depression Through Planned Obsolescence."

He pointed out that with the country in a depression, everyone was using manufactured goods much longer than they otherwise would have, and the lack of consumption was slowing down the economy. He suggested that the government set an expiration date on *everything,* from shoes to radios, and when it arrived, the expired items should be collected from the population, forcing them to buy new ones.

The same result could be accomplished by simply having the government go around and break all the windows every year. There is an underlying absurdity to the notion, like the old story commonly attributed to Mark Twain, which told of a small community on an island where the population "eked out a precarious livelihood by taking in each other's washing."

So where do we end up on the topic of obsolescence? Engineered obsolescence is clearly wasteful. Stylistic obsolescence isn't *necessarily* wasteful as long as out-of-fashion products build secondary markets. But it *is* wasteful when the items have no secondary life. And innovation obsolescence can't rightly be called waste by almost any definition.

Sixty Elements in Your Pocket

Smartphones provide ample fodder for those interested in waste in all its permutations. The newest and best models are launched at press events that rival Hollywood movie premieres; they cost a bundle, and people ooh and ahh over the latest, or fastest, or biggest, or smallest version—depending on what's fashionable at any given moment. Wait a few years, though, and you can't even give them away. They sit in drawers gathering dust.

What does the life cycle of a smartphone look like? How does it go from hero to zero—and how much of the process of creating these ubiquitous devices is lost to waste?

Like humans, smartphones begin and end as dust. To build one of these modern marvels, we start with the elements from the periodic table. In the case of your phone, it requires a *lot* of them. Sixty, in fact—which is more than can be found in your body.

Elements are substances that cannot be broken down into any simpler substance. Gold and carbon are just gold and carbon—

unlike, say, bronze, which is a mixture of copper and tin. There are 118 elements, of which 83 are both stable and non-radioactive.

Some of these elements are the so-called rare earth elements, which have become essential to our modern world. Despite the name, however, rare earth elements aren't actually rare. In fact, they're quite common. They're hard to obtain because their deposits aren't concentrated and they're difficult to mine and refine.

The most commonly used ones aren't all that expensive, despite how difficult it is to extract them. An ounce of lanthanum or cerium or samarium will set you back less than a quarter. Even the more expensive, heavy rare earth elements are still affordable—certainly compared to an element like plutonium (which, if you could even figure out a way to buy it, would cost you more than $100,000 an ounce).

In addition, the worldwide market for rare earths is rather small. The value of all the rare earth elements mined in a year is barely as much as the value of all of the copper mined in two weeks or the aluminum mined in a month.

So why are they such a big deal? Because they do very specific things in electronics and, for the most part, they have no substitutes. Think of baking a cake. It may call for three cups of flour and a tablespoon of baking powder. If you are short a tablespoon of flour, no big deal, but if you leave out the tablespoon of baking powder, well, you don't have a cake. Baking powder isn't expensive and it isn't rare, but it's essential to baking.

Rare earths have amazing powers. They make things glow brighter. They magnetize objects. They can be mixed with metals to make vastly stronger alloys. The kinds of products they enhance are critical; many are essential for modern living.

And our smartphones need most of them.

However, the process of mining and refining rare earths isn't easy and takes a long time. It's a nearly alchemical process involving acids, ovens, and proprietary processes. Imagine trying to extract a tablespoon of pepper that has been randomly spread through a pound of salt. That's the basic purity we're talking about with rare earth ore. And some of the processes take up to two years to run.

In addition to rare earths, a smartphone requires a number of other elements. By weight, the most common is aluminum. By value, it's gold. The rest of the elements all come together in their own idiosyncratic way from every corner and crevice of the planet.

Of course, they all have to be mined, a process that undergirds much of our modern world. Every person in a modern industrial economy like the United States, every single day of their lives, requires on average 20 pounds of sand, 15 pounds of coal, 3 gallons of petroleum, 1 pound of iron ore, 1 pound of salt, and 0.5 pounds of phosphate (which is used mostly in the fertilizers that grow the food we eat).

And that's nowhere near everything. According to the Minerals Education Coalition, in all, 100 pounds of ore must be extracted from the earth *every day of every year* for your entire life to support your modern lifestyle. And that's just in the United States.

Yet in the grand scheme of things, mining doesn't use much land, and what it does use can often be reclaimed. That's the good news. The bad news is that the amount of resources consumed doing all that mining, from water to fuel, is enormous, as is the resource consumption associated with all of the processing that must be done.

How much waste is there in mining? In a sense, it's all waste. In a world without waste, ingots of refined elements would be conveniently strewn about the surface of the earth in great abundance, ready to be picked up and used. But, of course, they aren't.

You can see part of the effective waste factor in their prices. Take gold, for instance. What do you think it would cost to buy a gold mine with 1 million ounces of proven gold reserves? First-order thinking might suggest that the value of that mine would be the price of gold—about $1,400 an ounce—times the number of ounces in that mine, so . . . $1.4 billion.

But in reality, because gold is so difficult to mine, you could buy the mine for a mere $50 million. That's right—gold mines are generally valued at $50 per ounce of gold in them.

If gold sells for $1,400 an ounce, purified, and the gold mine is only valued at $50 an ounce, then it's easy to see that the $1,350 difference isn't the value of the gold but the waste inherent in extracting and refining it.

But even that $50 an ounce is waste. The mine is worth $50 an ounce only because significant amounts of labor and risk went into determining that it was in fact a *proven* source of gold. In a world without waste we never would have had to expend the capital and take that risk to make the assessment. Not every hole gold miners dig has "pay dirt" in it.

The value of all the gold in a random 100-acre tract of land is essentially . . . zero. Gold is, in fact, worthless, in the sense that the minuscule flecks of gold that may or may not be in your back-yard aren't worth anything. You can't go to the bank and borrow money against the gold mining rights of your backyard.

In addition to the waste implicit in the price of the elements, there is the human toll as well. Many mines are inherently hazardous to both people and the environment. Child labor and even forced labor are used around the world. Can the human cost be eliminated or at least mitigated? Can't those who manufacture goods like our smartphones do something about these issues? Yes, but it's not as simple as that, since the supply chains that bring these elements together touch untold millions of people.

Imagine a small village in the developing world. It may have a mine that has no miners as employees. Instead, anyone can come to the mine, chip out ore, bring it to the surface, and be paid for its weight. That ore is sold by the truckload to an intermediary that sells it by the trainload to a refinery in a distant city that buys from scores of producers. The chemicals needed to refine the ore are likewise made in distant facilities that have their own supply chains across a dozen countries.

The elements produced from the ore are made into components which are then shipped to another facility where a different step in the manufacturing process occurs. Those unfinished goods are aggregated at a different place in perhaps a different part of the world to have another step in the manufacturing process performed on them.

If you are in the business of building millions upon millions of smartphones, the number of people and places involved dwarfs the imagination. You simply cannot effectively police every step of the process to make sure best practices are always used.

What you can do, however, is require those you do business with directly to follow certain guidelines, and require that they require their suppliers to do the same, all the way down the supply chain. Abuses are inevitable because there is almost always money to be made in the abuse business. But responsible manufacturers do surprise inspections of their suppliers, who are required to do likewise to theirs.

Could a smartphone be made that uses only ethically sourced materials built in a sustainable way? As it happens, Bas van Abel set out to do just that when he founded Fairphone. The goal was to make a phone with a long-lasting design, completely modular so that it can be easily repaired and upgraded, built under good working conditions out of materials that were sourced from sustainable and safe environments. Oh, and the phone needed to be easily recyclable.

Clearly, the idea had merit. Fairphone began as a crowdsourcing project and sold twenty-five thousand units before production even started. To date, the company has sold well over a hundred thousand phones and is even "mining" resources from old, discarded phones.

To its credit, Fairphone isn't just taking the easy path to building a "fair" phone. The easy path would be to deal only with well-regulated developed nations—by, for instance, purchasing materials from Australian mines instead of those in the Congo, where conditions are much murkier. Instead, Fairphone goes into the areas that are the most challenging and works with producers to improve conditions.

But even this practice is far from perfect. Van Abel admits, "The problem with that is you work with mines where there might be child labor. You work with mines where the working conditions are still not super-duper." He warns, however, that a more heavy-handed approach would end up backfiring. "If you go into a policing mode and tell them all what to do, they'll just show you what you want to see."

Even with the best of intentions, which van Abel clearly has, it's difficult to fully execute his vision.

The waste—both in materials and human costs—is significant just to collect the materials of our smartphones. To go into the waste involved in manufacturing would occupy an entire book, so in the interest of brevity, let's assume the new phone gets made and is delivered into the hands of an eager consumer.

Now let's jump forward in time a few years. That fancy new smartphone we meticulously built has reached the end of its useful life, at least as far as the person who bought it is concerned. What happens to the phone then?

Surely, given the enormous cost, difficulty, and waste involved in mining all of the elements used in the phone, the valuable material in the device can be quickly and efficiently recycled, right?

Not quite.

Why *aren't* smartphones recycled as much as, say, aluminum cans? At first glance, phones would seem to be ready-made for recycling; they are little containers of all kinds of conveniently ordered precious metals, all in one package. Throwing one out in the trash seems the epitome of waste, akin to a cartoon fat cat lighting a cigar with a hundred-dollar bill.

But this is where the challenge lies. You've probably heard the statistic that the value of all of the chemicals in the human body is something like $1.92. On a theoretical level, that makes sense to most; after all, we're made of relatively common stuff—carbon, hydrogen, oxygen, calcium, et cetera. Our smartphones, on the other hand, contain aluminum, gold, lithium, and all sorts of other materials.

But here's the bad news—the raw materials in your cellphone are worth even *less* than the raw materials in your body. Only about half as much, actually.

Let's look at the gold in a phone, which makes up half of the scrap value of the whole device. Worldwide, the average yield of primary gold production (that is, gold that originates in a mine) is 1 gram of gold per ton of ore. A gram is the weight of a paperclip. One ton of ore is about the size of a mini-fridge, the kind you might see in a college dorm room.

The kind of phone you think about recycling—say, an iPhone 6—weighs about 5 ounces. That means there are about 6,400 iPhones in a ton. According to metallurgist David Michaud of 911Metallurgy Corp., an iPhone 6 has 0.014 grams of gold in it. So our ton of iPhones contains about 90 grams of gold. So far, so good. The density of gold in an iPhone compares very favorably to what we might find in nature—90 grams of gold in a ton of iPhones versus 1 gram of gold in a ton of ore.

It's safe to say that if a gold prospector down in a mine were to come across a vein of pure iPhone, it would be cause for great celebration. Alas, the mother lode of iPhones is rarely found in nature.

The 3 ounces of gold you get from your ton of iPhones is worth $4,200 at $1,400 an ounce. This works out to about 65¢ a phone.

Here's the catch: The problem with recycling iPhones is not dissimilar to the problem with mining rare earths—they're distributed widely. Since melting down your iPhone at home isn't a current option, we'd have to figure out a way to get that device to a smelter for less than 65¢. You could hardly mail one for that much, let alone figure out a way to buy the unwanted iPhones.

But there's still hope, since there are other substances of value in the phone, such as copper and nickel. To get the amount of metals found in a 5-ounce iPhone from primary production would require about 2 pounds of each respective ore. At current prices, the nickel in the phone is worth about 3¢, and the copper 5¢. The iPhone 6 contains about 31 grams of aluminum as well. This metal is worth about $1,500 a ton, which makes the aluminum in the iPhone worth about a nickel. Beyond gold, copper, and nickel, as the saying goes, the juice simply isn't worth the squeeze. Other metals in the phone are used in such minute quantities that it's more efficient to mine high-grade ore than to extract from the low-grade "ore" of the phone.

So . . . 65¢ for the gold, and 13¢ for the other metals. It's difficult to make the numbers work.

Apple, however, is a bit more optimistic. In 2017, Apple launched a war against waste, hoping to one day be able to make its products from 100 percent recycled materials rather than using *any* primary production metals, and going so far as to launch a pilot program to build robots that can disassemble iPhones and other devices. Spoiler alert: They're not very close to achieving this goal.

The vast majority of discarded devices are still operational. So instead of shredding and melting down the phones or disassembling them with robots, what about reusing them?

Functional recycling happens to a certain point already. As previously noted, an early adopter who buys a $1,000 smartphone every two years generally doesn't toss it in the trash when deciding to buy a new phone. That phone is probably still worth at least $100, an amount most people don't cavalierly ignore. There are a number of ways that phone finds itself in the hands of a new owner.

But several years after it has been reused, passed along, and gone through a second or third lifecycle, the phone's value is down even more, perhaps to just $5 or $10. At this point, it isn't worth the trouble of selling the phone (transaction costs are a form of waste), but it still has enough value for someone to feel bad about chucking it in the trash.

Thus, it's likely to end up in the landfill known as the hall closet, or the ubiquitous junk drawer. Credible estimates suggest there are more than *two billion* cellphones stored unused in people's homes.

At the very end of its life cycle, no matter what happens, the phone is essentially worthless. What happens?

Marc Leff, president and cofounder of GRC Wireless, might enter the picture. His company buys *all* cellphones, regardless of condition. You ship it and they send you a check. They pay different prices for each phone, from a dime for an old flip phone up to $400 for the latest and greatest phones made.

We caught up with Leff and chatted with him about the ins and outs of his business, which he seems to regard as both a livelihood and a mission. The phones that arrive in large numbers every day are divided into two roughly equal groups. The first group is phones that still work and can be sold in secondary markets. As Leff tells us, the goal is to get five or six life cycles out of a phone across its useful life, which is fifteen years. In fifteen years, the finest and best smartphones of today will likely be worth nothing to anyone in the world. And this half of his business is the lucrative part.

The other half of the phones, the ones that Marc pays as little as a dime for, are largely a money-losing or break-even proposition. They are subdivided into two groups. The first are disassembled for parts, usually in less-developed countries. These are the phones old enough to not have much value themselves but which are still widely used somewhere. There is no supply chain for parts for these older phones, so the folks who disassemble them are providing a useful service and making a living doing it, ultimately reducing the waste that would occur with a phone that is perfectly fine except for, say, a broken pin on the charging port.

The final group, the old phones that are barely worth a dime, are sent off to smelters to have the "easy" metals, such as gold and aluminum, stripped from them.

Eventually, no matter how many life cycles you get from a phone, someday it will be worthless. And when that happens, unless there is major technological change that allows for recycling at home, phones won't get recycled, since it isn't even worth the postage to send them to someone (except perhaps a container-load at a time).

It's a paradox. On its own, one phone is essentially worthless. But in a given year, the world makes 1.5 billion smartphones and, sooner or later, those phones will have to be dealt with. In aggregate, now we are talking 22 tons of gold, and proportional amounts of all the rest of the raw materials. And this is just one single category of device, and a small one at that. Add in all the computer screens, TVs, laptops, and printers, and by one estimate, as far as gold is concerned, 10 percent of all new production could be replaced by recycling.

How do we keep this waste from occurring? The problem may seem intractable, but what if there's an easy answer?

There's an economics concept called "internalizing externalities." This practice charges the external costs of an action to the person or entity inflicting them. For instance, if Company X pollutes a river, then the damage done to the environment should be assigned a value, and whatever that value is should be levied in the form of tax on that company. Doing so forces the company to take into account the *total* impact of its actions, not just the ones it pays for directly. This method of taxation is also arguably the only form of assessment that can improve upon the theoretical efficiency of a free market.

If we can determine that a cellphone in a landfill inflicts $10 in damage to the environment, then we can enact a deposit system. When you buy a phone, you pay an extra $10 for it, up front. When the useful life of that phone is up, whoever owns it has two choices: They can toss it into the landfill, inflicting $10 in damage, or they can turn it in and get the $10 deposit back. Several states do this with beverage bottles: You pay a dime extra when you buy the drink, and someone gets a dime for returning the empty bottle. That's the same idea here.

If the societal cost of a cellphone in a landfill were zero, then the whole system would be reasonably waste-free. If a phone costs $1,000 today and fifteen years later it's worth a dime, 99.99 percent of the value of the phone was consumed. In a perfect world, if you could magically wave a wand and separate the phone into its core elements, they're only worth a dollar. Even if our $1,000 phone ends up being worth $1 and we throw it away, then we've still gotten virtually all of the value out of the phone.

But the societal cost of a cellphone in a landfill isn't zero. So what is it? We don't actually know, but we do know that smartphones are made of some pretty toxic stuff. There's arsenic in many of them, as well as lead and mercury. Less well known, but just about as toxic, are cadmium, chlorine, bromine, and lithium. The trend, however, is to use less of these substances. Apple touts in its iPhone X Environmental Report that the device features "arsenic-free display glass" and is mercury-free, PVC-free, beryllium-free, and free of brominated flame retardants.

In a sense, losing sleep over smartphones going into landfills misses a much larger problem. The combined weight of every phone manufactured last year is about 250,000 tons, but the amount of electronic waste the world produces each year is about *50 million* tons. That means that if you took every phone made this year and dumped them straight into a landfill, the resulting increase in the world's electronic waste would be a rounding error.

That 50 million tons of electronic waste works out to about 15 pounds a person. However, if you live in the developed world, you threw out more than twice that. You have to throw out a lot of phones to equal the weight of that microwave oven that got trashed. Your phone may have a tiny bit of lead in it, but that big TV sitting in your garage has at least 6 pounds of it.

In the United States, electronic waste accounts for just 2 percent of landfill volume, but that 2 percent accounts for 70 percent of all of the toxic substances in landfills. With recycling rates of electronic waste hovering around 20 percent, this is a problem that will only get worse.

Will we ever reach a point where it makes sense to mine landfills for the gold in them? Perhaps. The best assumption we can make suggests there are about 2 grams of gold per ton of landfill detritus—twice what typical primary production might yield. But the landfill ore is far more toxic than gold ore, so it may not make sense.

For those of you with a well-developed sense of the macabre, perhaps you're wondering about the ore content of cemeteries. If you were to mine the top 6 feet of a cemetery, the yield would be about 0.25 gram of gold per ton of dirt, given reasonable assumptions about what percentage of people are buried with their jewelry, medical devices, or gold teeth. Luckily for the dead, that is pretty low-quality ore.

Is All That Glitters Waste?

Gold has always had a special place in our hearts.

But why? Its allure seems purely aesthetic. Sure, it's pretty, and it never rusts or tarnishes. But otherwise it has no nutritional value, is a poor material for constructing shelter, and isn't useful for making tools or weapons.

Or perhaps we love gold because it's rare. For whatever reason, a great deal of energy, money, and time is invested in finding and digging up a metal that often just sits dormant in a vault somewhere. Is that waste?

The answer hinges on the question of value. When talking about gold, mining executives are fond of sayings like "An ounce of gold can always buy a man a fine suit." And that's more or less true. In one respect, the price of gold never really changes; the price of everything else does. A century ago, U.S.-minted $20 gold pieces each contained 1 ounce of gold. If at that time you wanted to purchase a custom-made men's suit, it cost about $20, or 1 ounce of gold. Today, that same ounce of gold will cost you $1,400, roughly the same price you'd pay for the same suit.

In 1915, the average cost of a house was $3,200, or 160 ounces of gold. In 2019, the average price of a home in the United States was about $225,000. Divide that by $1,400 an ounce, and that house costs 160 ounces of gold.

So the prices of houses, suits, and much else haven't changed when priced in gold. Other precious metals behave in much the same way: The price of a gallon of gas has hovered around the value of the silver in one quarter for the better part of a century. When silver coins circulated in the United States, a gallon of gas cost roughly 25¢. Sell that same antiquated quarter today for its silver content and you get roughly enough money to buy a gallon of gas.

Whether gold's appeal lies in aesthetics or value, our interest in it is so strong that we have mined it continuously for thousands of years. Archaeologists have found gold objects in present-day Bulgaria that are at least six thousand years old. Gold chains were being worn nearly five thousand years ago in ancient Ur, a city in today's Iraq. The Egyptians believed it was a heavenly metal and associated it with the sun god, Ra. In Hindu myth, gold is the soul of the world. And to the ancient Aztecs, it was *teocuitlatl,* literally "the excrement of the gods."

Lest we be too smug about our own enlightened age, today's scientists aren't even in agreement on how gold and the other heavy elements came into being (though few current theories involve the digestive tracts of deities).

The reverence we reserve for gold is evident when looking at our idioms around it. It represents achievement when you "go for the gold"; excellence when we talk about something being the "gold standard"; purity when we refer to someone as having a "heart of gold"; opportunity when a person is "sitting on a gold mine"; and wealth when something is "worth its weight in gold."

Gold is immensely dense; a quantity of it sufficient to fit in the bed of an average pickup truck would be worth more than $60 million at today's prices. The total amount that has been mined throughout human history is, interestingly, a point of hot contention for a couple of reasons. There is speculation that many countries either overreport their gold reserves or underreport them for various geopolitical reasons. Further, gold gets reused time and time again. If you buy a gold watch today, almost certainly it contains some gold atoms that were used to adorn the pharaohs. However, we have a pretty good range for the total amount of gold that has been mined over the course of history: On the low end, enough to fill an Olympic-sized swimming pool. On the high end, enough for three such pools.

With regard to scarcity, in the earth's crust about one atom in a billion is gold. However, if it were distributed evenly, it's unlikely we would ever notice it or have been able to mine it. Luckily, it isn't quite that dispersed and sometimes appears on the surface as reasonably pure nuggets. More often it's found in lode deposits, or gold veins—when molten gold from deep within the earth is pushed up into the cracks in rock. Further, since gold is heavy, it collects on the bottoms of rivers. Over geological time, a river bottom will develop discernable strata, or layers, of rocks, with one of those layers sporting a relatively high gold content.

Because of the range of ways that we find gold, the yield of a ton of gold ore varies immensely, from less than 1 gram of gold per ton of ore to nearly 50 grams per ton. You might suppose the latter type of ore is preferred to the lower grade, but it's not quite that straightforward. Gold ore in Nevada, for instance, is low grade, but it can be had in great volume using just surface equipment for extraction, whereas higher-grade ore may be deep within the earth and harder to get to.

Regardless of how we get gold ore, we have to refine it. There are a number of techniques that vary based on the composition of the ore itself, but we can look at one as representative of the whole. In this process, the first step is to pulverize the ore and mix it with water and cyanide. The mixture is infused with oxygen, which reacts with the cyanide to dissolve the gold into the water. Then the water is drained off and zinc powder is added to it, which solidifies the gold again. Now you have a zinc and gold mixture and a bunch of cyanide water. Next, another six elements are added to the zinc and gold mixture to form a substance called flux, which is then heated to 3,000 degrees Fahrenheit. It bakes for a couple of hours, in which time the gold sinks to the bottom and everything else, called slag, floats on top. The slag is poured out and analyzed to make sure most of the gold is gone. Then the remaining gold is poured into a mold to form an ingot.

The amount of waste in the mining process is easy to see. Huge amounts of capital are spent finding mines and developing them. Roads have to be put in. Equipment has to be made. Imagine the trial and error in developing the one process described above. Human capital is spent mining. Vast amounts of fossil fuels are consumed at each stage of the process. And cyanide isn't exactly harmless. The environmental impact of gold production is large. And the refining process is inherently wasteful, because some gold is lost at every step.

All this effort, all this energy, all this expense, is traded for a substance in which 90 percent of all production is used, in roughly equal parts, for body adornment or as a store of wealth. Only a tenth is put to practical use, most of which finds its way into electronics, and most of that . . . ends up in landfills.

Is all this effort a waste? People do value it, so in one sense it isn't. However, many are left scratching their head as to why we value it so much. Warren Buffett frequently ridicules the idea of buying gold as an investment. He points out that if you wanted to buy all the gold in the world, it would cost you $10 trillion. Alternatively, you could use that money to buy all 400 million acres of cropland in the United States as well as sixteen Exxon Mobils— and still have a trillion dollars in pocket money left over. He then says:

A century from now the 400 million acres of farmland will have produced staggering amounts of corn, wheat, cotton, and other crops—and will continue to produce that valuable bounty, whatever the currency may be. Exxon Mobil will probably have delivered trillions of dollars in dividends to its owners and will also hold assets worth many more trillions (and, remember, you get 16 Exxons). The 170,000 tons of gold will be unchanged in size and still incapable of producing anything.

Buffett's point is true. As indicated at the beginning of this chapter, an ounce of gold bought one men's suit a hundred years ago, and it still only buys one men's suit today. It is a store of value, not an income-producing asset.

When the world was on the gold standard, all official currency was backed dollar for dollar by gold. If a country wanted to grow its money supply, it had to dig a hole in the ground and look for gold. Great amounts of energy were expended to pull up a metal to sit in a vault to back a currency. While that practice may seem wasteful from one viewpoint, to any country that has experienced hyperinflation, there's a reassuring stability to that.

Occasionally gold can be inflationary, such as when the Spaniards brought great hoards of the stuff back from the New World. But generally speaking, its relative scarcity and the difficulty of finding it are its actual virtues as a source of money.

Of course, all that could change with technology. The asteroid 16 Psyche, which lies between Mars and Jupiter, is estimated to contain metal (including gold, but also platinum and nickel) worth around $700 *quintillion*. What's more, the asteroid is made nearly entirely of metal. If that could ever be harvested economically, it could force the price of those metals down to a point where, like in the legendary city of El Dorado, the streets could literally be paved in gold.

Should some other calamity befall our civilization, we may find our paper money worthless. But unless the supply of gold increases dramatically, it will almost certainly retain its value. As such, while humanity's use of gold as a store of wealth may be far from optimal, it is not, strictly speaking, wasteful.

Mine Locally, Smelt Globally

In a later chapter, we explore the locavore movement and determine an answer to the question of whether the amount of waste inherent in food consumption is a function of the distance the food travels to its final destination. This analysis can also be applied to metals.

The produce you consume travels on average 1,500 miles from where it is grown to where you consume it. Despite that distance, the energy used to transport produce is only a small fraction of the overall energy used in growing the food.

Aluminum ore, on the other hand, travels twice that distance on average, from where it's mined to where it's refined. Given that ore is incredibly heavy and shipped by sea on diesel-burning, carbon-spewing container ships, surely that distance *must* represent a tremendous amount of waste, right? Not necessarily.

Aluminum comes from a sedimentary rock called bauxite, which contains a great deal of aluminum—enough that 5 tons of bauxite will eventually yield 1 ton of pure aluminum. In contrast, 5 tons of gold ore will yield a mere 2.5 ounces of gold, and the weight of your house in copper ore would need to be dug up just to get enough for the electrical wiring for that one structure. Compared to all of that, bauxite is relatively pure stuff.

While bauxite contains a lot of aluminum, it's tricky to coax the metal out. Doing so involves two processes. The first is known as the Bayer process and results in aluminum oxide, or alumina, an intermediate product that must be further refined. To turn alumina into aluminum, the most common method is the Hall-Héroult (HH) process.

Bauxite is heavy. A carry-on-sized suitcase full of it weighs about 60 pounds, and alumina is heavier still: that same suitcase full of alumina weighs about 180 pounds. Yet these raw materials travel nearly a tenth of the earth's circumference from where they're mined to where they're ultimately transformed into metal. How can adding this much distance to the process *not* be wasteful?

It comes down to energy. Each stage of aluminum production requires a large amount of power, and the HH process demands electricity specifically—so much of it that industry insiders colloquially refer to the small bricks of newly produced aluminum as "packaged electricity."

Smelting a ton of aluminum—enough to make sixty-four thousand beverage cans—requires about 15,000 kWh of electricity. That's more energy than one home in the United States uses *in an entire year*.

Aluminum sells for about $1,500 a ton. The average residential customer in the United States pays around a dime per kWh for electricity. If we do that math, we can estimate that it requires about $1,500 worth of electricity at residential prices to make $1,500 worth of aluminum. Refining bauxite in your basement wouldn't be a great idea. In fact, it would be quite wasteful—the residential electrical grid isn't designed to deliver that much electricity to the average house. Upgrading local transmission lines to get you the power you would need to smelt would be extremely inefficient.

How much more efficient are commercial producers of aluminum compared to our theoretical home smelter? In general, to produce the metal profitably, a refiner must keep its energy costs down to between 33 percent and 40 percent of the selling price of the metal. Clearly, refiners can't pay full retail for their electricity.

Aluminum producers have to buy electricity for less than half of what residential customers pay. In China, the world's biggest aluminum producer by far, refineries pay a nickel per kWh as a result of heavy subsidies, which means they have energy costs of around $700 for a ton of aluminum.

In the United States, production of aluminum from ore has dropped precipitously since 1998, when America was the world's largest producer. Around the turn of the millennium, there were twenty-three operational aluminum smelting plants; twenty years later, only a handful remain. As in China, the plants in the United States that have survived tend to be highly subsidized; Alcoa's Massena West smelter in upstate New York has received nearly $70 million in subsidies from the state over the past three years, even as the global price of aluminum has increased by nearly a third.

Where production isn't subsidized, aluminum producers must be creative. Just as Willie Sutton robbed banks because "that's where the money is," aluminum smelting is generally done in places where there are sources of inexpensive electricity.

Iceland, for instance, with its vast capacity for geothermal power, is home to three aluminum refineries that consume 70 percent of all of the electricity generated on the island. It's that lower-priced electricity that makes it worthwhile to ship bauxite there—in some cases from as far as Australia. The distance from Sydney to Reykjavik is just over 12,000 miles—they are literally halfway around the world from each other.

Just how much efficiency is there to be gained by shipping bauxite to a refinery as opposed to refining on-site? Imagine two options. In one, all bauxite refineries are built next to bauxite mines. In the second option, all bauxite refineries are built in places with cheap clean energy and the ore is shipped in. How do those options compare? Benjamin McLellan did the math in his fascinating paper "Optimizing Location of Bulk Metallic Minerals Processing Based on Greenhouse Gas Avoidance": "With current grid electricity, localization of aluminum and alumina production in bauxite-producing countries, would result in an overall increase in emissions by roughly 14 percent. Relocating to Norway would result in a reduction over current emissions of approximately 44 percent."

This is a big deal. An increase in emissions of 14 percent equates to about 90 million more tons of CO_2 per year, a number the EPA equates to all the energy used by 10 million U.S. households for a year. That's a number larger than all the households in Texas. (And keep in mind that this increase is a *net* increase; McLellan even subtracts the carbon that comes from sending container ships around the world.) This reduction in waste would result from the fact that not all sources of electricity are equally efficient.

Given how we currently generate electricity, there's literally no place on the planet that's so remote that it isn't dramatically more efficient to ship the ore to places with sustainable clean energy than it would be to process it on-site.

Processing bauxite uses so much electricity that great financial gain is the reward for any company that can figure out a new way to do it. One promising lead is a joint venture by aluminum companies Alcoa and Rio Tinto along with Apple to commercialize a new zero-carbon method for smelting aluminum. Others are trying to apply technology in different ways. One initiative seeks to substantially lower the temperature at which the smelting process occurs, which in turn lowers energy usage. An even more ambitious effort seeks to replace the current smelting process with a new chemical-based one.

In the future, *Star Trek*–style transporter technology could be used to get ore to where the cheapest and cleanest energy was available, and then to deliver the finished aluminum to where it was needed. Until that happens, however, the conclusion is clear (although it seems counterintuitive): More often than not—at least in the case of making aluminum from bauxite—it makes sense to send rocks on the slow boat from Sydney to Reykjavik.

Waste at 30,000 Feet

The widespread adoption of safe and economical commercial aviation has contributed to a massive reduction both in time wasted in travel and lives wasted in traffic deaths. But is flying itself wasteful?

The short answer is yes. So much so that it's becoming stigmatized in certain circles, mainly due to its emissions of CO_2. No one disagrees that flying produces a lot of carbon; every mile flown burns 2 ounces of fuel per passenger. And as we'll see in the chapter on plants, because carbon binds with oxygen to make CO_2, each passenger mile flown produces more than 6 ounces of CO_2. A thousand-mile flight—one that lasts around two hours—produces about 400 pounds of CO_2 per passenger. Given that the average American produces 36,000 pounds of CO_2 a year, it doesn't take many flights to make up a significant portion of our carbon footprint. And flying is only projected to increase.

To fully explore the question of waste, we should compare flight to its closest alternative: driving. How do the carbon emissions from aircraft compare to those from automobiles? A gallon of gasoline weighs about 6 pounds and produces about 20 pounds of CO_2. If your car gets 20 mpg, then it's emitting 1 pound of CO_2 per mile. If you drive 1,000 miles in a car by yourself, you emit about 2.5 times what the flight would emit. Stated another way, planes get about 50 mpg per passenger. To get geeky about it, though, driving and flying aren't perfect substitutes. Seldom do you fly down to the corner store for a gallon of milk or drive to Hawaii. Additionally, a typical business traveler in the United States might drive about 10,000 miles a year but fly 20,000. If airplanes didn't exist, that business traveler probably wouldn't drive 20,000 additional miles.

But the waste in flight goes beyond carbon emissions. Let's look at what passengers waste. Imagine all the uneaten food, half-eaten food, little wine bottles, half-used amenity kits, credit card application forms, cheap earbuds, empty cups, and the like. Together, in 2016, this detritus totaled about 11 billion pounds, or about 2.5 pounds per passenger per flight. (Okay, that number includes bathroom waste, which isn't fair. But assuming half the people on the plane use the toilet once, there's still 2 pounds of waste per person per flight.)

Airlines are aware of this inefficiency—every spare ounce that flies costs them money—and are making efforts to reduce it: changing how uneaten food is disposed of, donating unused amenity kit contents to shelters, using compostable cups, and a score of other initiatives. There's a growing awareness that this sort of waste is a problem that can at least be partially solved without great effort or expense.

Another form of waste found in air travel has to do with routes and gates. There is a finite number of airport gates in the world, and thus a finite number of flights that can be accommodated. Some of these flights are inevitably oversold, while others fly less than full. Both of these are problems that airlines wish they could solve.

You may remember an incident a couple of years back when a

man was dragged off a United flight to make room for another passenger, as the flight was oversold. If you were that other passenger, you probably would regard United as having the best customer service on the planet. But everyone else shuddered at this story, imagining they were the removed passenger, who in this case happened to be a doctor who had appointments with patients the following day.

This single incident caused several airlines to increase the compensation they pay to have people voluntarily leave overbooked flights from a few hundred dollars to an eye-popping $10,000. Seen through the eyes of economic theory, this is the right way to handle the problem. If you offer passengers $5,000 to sell their tickets and none do, then you need to up your offer (not drag a passenger off the flight).

Undersold flights are also a source of angst for airlines. Every so often someone posts a photo online of them being the only person on a long-haul commercial flight. While most of us think having the run of the plane would be fun, CEOs of airlines wince when they see these stories.

Artificial intelligence is the most likely way to eliminate, or at least mitigate, the waste here. There are 100,000 commercial flights a day, each of which generates a wealth of data about the number of passengers, how much they paid for their tickets, and so forth. This data is, and will increasingly be, used to refine the flights that are offered. AI should allow more planes to fly full while lowering the average price of a ticket.

Airlines are making significant progress in reducing waste. Virgin Atlantic determined that removing just a pound of weight from every plane would save them 14,000 gallons of fuel a year. This calculation led them to make their drinking glasses thinner and offer lighter desserts. American Airlines managed to persuade the Federal Aviation Administration to let them ditch the 35-pound paper flight manuals in favor of iPad copies, because the fuel savings would be $100,000 *a month* across their fleet. American also figured out that the fuel savings to be had by ditching the SkyMall catalogs from their fleet would save another $30,000 a month in fuel. United saved nearly $30,000 a month in fuel costs by printing its in-flight magazine on thinner paper. In the airline industry there's a long history of these sorts of small changes having big impacts. You've probably heard about how, back in 1987, American found that it could save $40,000 a year by removing a single olive from everyone's dinner salad. Did removing that olive reduce waste? It depends on how much you value an extra olive.

Industrial design studio PriestmanGoode has suggested that replacing single-use plastics with water fountains on long-haul flights could result in a staggering reduction in the amount of waste. On just one route—Singapore Airlines' eighteen-hour nonstop flight between Singapore and Newark, New Jersey—the firm claims that fountains could reduce the number of plastic bottles used by as many as 3,400 *per flight.*

When looking at ways to reduce waste in aviation, however, we need to be mindful of the phenomenon we saw in the chapter on recycling. When consumption becomes more efficient and less expensive, people naturally increase their consumption. If flying continues to become less and less expensive, people will fly more. Will flying more result in a net increase in waste? The answer to that question requires too many subjective value judgments for this book to give a definitive answer.

If You Don't Like It, Just . . . Return It

Given that clothing is something everyone needs, the fact that in an $80 trillion world economy we only spend 3¢ out of every dollar on clothes seems low, particularly since fashion is an important element of popular culture.

On a per capita basis, that means that we spend about $300 per person per year on clothing. Of course, that number is a bit misleading; those in richer countries spend vastly more than those in the developing world. In the wealthy West, people buy so much clothing that the amount purchased annually but *never actually worn* is more than the $300 average.

Clothing consumption is driven by low costs and rapidly changing tastes. We think little of purchasing new clothes, then purchasing more when the last batch no longer suits our fancy.

This view of what we wear as disposable is new. At most times in history, people were stuck with the clothes they had for much longer than a season. This is why, for instance, the progression of belts signifying skill level in karate goes from light to dark, so that as the student advances they could redye the same belt in ever-darker shades.

Even today, men's dress shirts often feature collars and cuffs in a different color from the rest of the shirt, recalling an earlier time when men would have those parts of shirts replaced when they became soiled and worn, saving the rest of the garment in the process.

Previous generations' frugality in attire can be seen even more clearly in a legendary, and entirely true, story from the 1920s. At that time Asa T. Bales noticed that women were repurposing the cotton bags that animal feed was packaged in to make clothes. He patented the idea of printing patterns on those fabrics as a kind of differentiator for feed manufacturers, and feed sack dresses are commonly seen in reports of clothes from the era.

Contrast this previous expense and scarcity with practices in the modern clothing industry, and you can see that there's been an explosion in waste. Readers may recall the controversy from 2013 when a manager at fashion-forward retailer Abercrombie and Fitch said that they burned remaindered and irregular inventory rather than donating it to the needy, implying that the company's brand would be harmed if poor people were seen wearing the clothes. Today Abercrombie's website says that the company has since "partnered with World Vision, a global humanitarian organization that provides clothing to families and children in need around the world. We donate as much as we can, but like many retailers, there are certain goods we cannot give away due to regulations and damages."

Abercrombie is certainly not alone in being concerned about who might end up wearing its castoffs; maintaining exclusivity can be something of an obsession for premium brands. A piece in *The Sunday Times* reported that luxury retailer Burberry had burned more than $100 million worth of its own goods over the past five years. As the article notes, "Designer labels, it is claimed, do not want their products to be worn by the 'wrong people' after emerging on to 'grey markets' at knockdown prices."

Today we are so awash in clothing that it staggers the imagination. By his own admission, Justin Bieber doesn't wear the same underwear twice because he is given so much free underwear by Calvin Klein. George Clooney is rumored to have his socks laundered and donated to a homeless shelter after a single wearing.

Is it waste to wear a garment once? Well, strictly speaking, by the standards of this book, it depends on the preference of the person. If twice-worn socks burn your feet in the same way that the elven rope burned Gollum's neck in *The Lord of the Rings*, then you have little choice but to throw out socks daily, or even hourly if it comes to that. But, practically speaking, the useful life of those socks is far longer than a day.

At some point, the question is not one of utility but rather one of fashion. The notion of fashion, and the exclusivity it implies, is hardly new. Styles have always changed from year to year, and fashionistas all over the world have frantically tried to keep up. Thomas Jefferson advocated that the men and women of his time should, "in matters of style, swim with the current." Shakespeare opined about one of his characters that "the soul of this man is his clothes." Roman poet Ovid weighed in with pages and pages of fashion advice, as did Voltaire and Ben Franklin.

Of course, there have been contrarians; one such writer was Quentin Crisp, who remarked, "Fashion is what you adopt when you don't know who you are." But for the most part, people follow the hemline du jour, and do so enthusiastically.

The importance of fashion is front and center in the 1925 classic *The Great Gatsby*, when the title character, wanting to impress the object of his affection with his wealth, opens his wardrobe: "[Gatsby] opened for us two hulking patent cabinets which held his massed suits and dressing-gowns and ties, and his shirts, piled like bricks in stacks a dozen high. 'I've got a man in England who buys me clothes. He sends over a selection of things at the beginning of each season, spring and fall.'" This trend was well under way before the fictional Gatsby was even born.

What's relatively new is the democratization of fashion, which dates back to at least Eli Whitney, whose gin drove down the price of cotton. Suddenly, almost everyone could afford new clothes before the old ones wore out. Automation and globalization have fed that beast, forcing prices down further and further until we arrive at an era of fast fashion where hyper-trendy, low-quality clothing is purchased and worn, by one tally, just five times before being discarded.

Evidence beyond the anecdotal exists that social media is exacerbating the trend, as some people don't want to post photos of themselves "repeating" an outfit. In this world, the most biting insult someone can leave on a photo is "I always love you in that dress." Contrast this ethos to actor Daniel Radcliffe's practice of wearing the exact same outfit for months at a time, ostensibly so paparazzi can't sell photos of him since they look like they could have been taken weeks ago.

How many sets of clothing did previous generations use? In the first half of the twentieth century, Americans spent 13 percent of their income on apparel. The average family made $750 in 1900, meaning $100 of it went toward clothing—that's a huge proportion. Now we spend just 3 percent of our income on apparel and, by number of articles, buy about six times as many because the prices have plummeted.

Going back a bit further, you may have noticed that older homes from around the nineteenth century usually lack two architectural features found in all modern homes: bathroom counters and bedroom closets. People used specific furniture for these items—the vanity and the wardrobe, respectively. The small size of antique wardrobes compared to today's walk-in closets is telling. Antique furniture was generally nowhere near large enough to accommodate anything like the amount of clothes we have today. The wardrobes that survive to this day would have been for those with middle- and upper-tier incomes. The walk-in closet, now common in American homes, is on average the same size as a spare bedroom in 1950. The first walk-in closets date only from about forty years ago, by which time globalization had dramatically reduced the cost of clothing.

Dialing the clock back further, consider the writings of the great French historian Fernand Braudel. His three-volume work *Civilization and Capitalism, 15th–18th Century,* explores the economic conditions of the average people of this period, partially as a refutation of Marx's economic view of history. He writes extensively about clothing, pointing out that while people then didn't have the variety of clothing we have today, the image of the peasant with just one set of clothing probably isn't true. He writes that in the 1700s in Sardinia, the custom was to wear the same shirt for a year when mourning, implying that more frequent changes were the norm. He quotes a contemporary observer in England remarking that the decline in disease over prior centuries was due to the peasants having more sets of clothing.

The world is so awash in clothing in the modern era that the average person worldwide adds *a garment a month* to their wardrobe, a rate twice what it was just fifteen years ago. This practice implies that a garment a month per capita will eventually find its way out of a wardrobe, and it will most likely end up in a landfill.

The fact that the average American throws out about 2 pounds of clothing per week lends credence to this theory. Indeed, according to the EPA, 60 percent of it is landfilled. Another 20 percent is burned for fuel. The final 20 percent gets recycled.

These numbers seem incongruent with our real-world experiences. Most people don't throw 60 percent of their clothes into the trash, burn 20 percent for fuel, and recycle the other 20 percent. Apart from ruined garments and underclothes, many Americans' default method of disposal is to donate the clothes to a charity, such as a thrift store. It would be nice to think that the clothes we donate end up going to the needy. It almost makes buying a new outfit an act of charity, knowing that in a year you will pass it along and someone less fortunate will be the beneficiary of your benevolence.

But that's not what happens. The world is awash in clothes, and the amount of donated clothes that end up being worn by someone else is barely over 10 percent. The vast majority is ultimately landfilled, and an additional good portion is shredded for use as insulation or turned into shop rags.

Local charities get first bite at the small portion that remains, as they select the donated items they think have the best chance of selling. More than a few of these items will actually be new, with original tags attached. Eagle-eyed spotters snatch up leather, cashmere, premium brands, and other desired items, price them, and place them in stores. But tags are almost always color-coded by date, so if an item doesn't sell in a set period of time, perhaps a month, it gets thrown out to make room for newer donations.

The rest of the clothes are sent overseas and sold there by the bale. Some people are critical of this entire process, calling it a form of colonialism that not only undermines local textile markets but also has a negative impact on indigenous forms of dress, biasing tastes toward the logo-emblazoned T-shirts and blue jeans of the West.

Another school of thought rejects this reasoning, saying it overlooks all the jobs that the influx of cheap clothing creates. Employees are needed to mend them, alter them, launder them, and ultimately sell them. In addition, the money that nonprofits make selling clothing to exporters can be used for good as well. Passions run deep on this issue, and several countries in East Africa are in the process of banning—or at least heavily taxing—the import of secondhand clothes and shoes in order to protect the local textile industry.

This revolving door of clothing, which sees copious garments landfilled before their useful life is over—or even before they've been worn—is made possible because the modern economy is incredibly efficient at making inexpensive clothes.

Clothes are so cheap to make that it can make sense for manufacturers to make twice as many units as they know they'll need. Sound crazy? Consider the producers who routinely make multiple versions of their products—shirts, jackets, caps, sweatshirts, all of it—for the two teams that reach the Super Bowl, declaring each to be the winner, knowing that when the big game is over half will be disposed of, usually to impoverished parts of the world. This practice isn't confined to football; all manner of professional events do this. The euphoria that prompts a fan to buy an article of clothing celebrating their team's victory is ephemeral, wearing off in days, even hours. The half of the merchandise that's lost is just the cost of doing business. And even the winning team's clothes have a limited fashion life.

The fast-fashion phenomenon has significant ripple effects beyond the disposal of the clothes themselves. Environmentally, the clothing industry is pretty toxic when you consider the impact of the chemicals required, such as bleach and dyes, as well as the alchemy that goes into making synthetic fibers. But natural fibers aren't exempt from criticism, either; the production of cotton and linen is both land- and water-intensive. According to an in-depth study of the environmental impact of a pair of Levi's 501 jeans, Levi Strauss & Co. determined that producing the jeans required over 600 gallons of water and emitted over 40 pounds of CO_2. A report by the sustainability consulting company Quantis concluded that the apparel industry accounts for well over 6 percent of human-made global CO_2 emissions, even though apparel is only 3 percent of the world economy. (To put this figure in perspective, the apparel industry's emissions are roughly triple that of the entire aviation industry.)

Interestingly, the amount of resources we spend on clothing *after* it's been made is often more than what it takes to make it. Clothing, after all, requires maintenance. We frequently bathe it in water that has gone through water treatment plants and been warmed with fossil fuels. Is there waste here? Do we wash our clothes too often, wasting valuable resources? Levi's president Chip Bergh thinks so, at least with regard to jeans. He often sports a ten-year-old pair of his company's product that he says has never seen the inside of a washing machine. Anderson Cooper has similar views, once saying that he only washes his jeans a few times a year and does so by wearing them into the shower and then hanging them up to dry.

It's nearly impossible to determine the unintended consequences of disposable fashion. We may, however, see some glimpses of the impact by examining other historical eras. According to historian Barbara Tuchman, we've been in this situation before. It was the mid-1300s, and perhaps a third of Europe had just died from the Black Death. People had to find a purpose for the surplus clothing that had previously belonged to people who died.

Papermaking began in Europe around this time, using this excess clothing as raw materials. New paper was produced in abundance, unlike the expensive and limited parchment that it replaced. A century later, Gutenberg brought Europe movable type, which would've been irrelevant in a world of parchment or expensive, handmade paper. It isn't *too* much of a stretch to say that excess clothing was one of the most important ingredients in the creation of the modern world. Unfortunately, at present, no new "killer app" for our excess clothing has materialized.

Technology makes figuring out the most appropriate uses for

discarded garments more efficient. It will soon be possible to embed computer-readable electronic tags in every garment made, as such chips are falling in price dramatically and will soon cost less than a cent. If there was a way to know more about what was in any given piece of clothing, machines could sort it all. Clothing that was 100 percent cotton could be remade into shop rags, while synthetic fibers could be turned into fuel.

It's unclear if market forces will naturally advance this vision; doing so universally might require a government mandate.

Ultimately, as we move along the continuum toward less waste, our relationship with clothing will change. Perhaps our garments will have settings that change their colors and patterns, allowing a couple of shirts to go a long way. In the interim, services that allow people to share ownership of clothing, in the same way that many rent formalwear, might fill the need for variety. Eventually, clothes won't wear out, and every pair of socks will feel like a new pair each time it's put on. George Clooney should be pleased.

Returned Goods

When online ordering was new, there was a general worry about buying goods that you hadn't seen, handled, or tried on. That concern seems almost provincial today. But in the 1990s it was a real thing, right up there with worries about entering your credit card number online (and Y2K). Who would do such a crazy thing?

To mitigate this concern, online sellers took a page (almost literally) from catalog sellers of old. More than a century ago, Sears made this promise to its customers in its 1918 catalog:

> WE GUARANTEE that each and every item in this catalog is exactly as described and illustrated.
>
> We guarantee that any item purchased from us will satisfy you perfectly, that it will give you the service you have the right to expect; that it represents full value for the price you pay.
>
> If for any reason whatever you are dissatisfied with any article purchased from us, we expect you to return it to us at our expense.

We will then exchange it for exactly what you want, or will return your money, including any transportation charges you have paid.

As liberal return policies at online retailers became the norm, brick-and-mortar stores adopted similar policies. Over time, retailers would one-up each other with the length of their return periods. That's how we got to today's world of easy returns.

But it didn't stop there. Returns weren't just easy. Along came free shipping for returns. Or, if you preferred, there were local retailers where you could drop off your returned item. In some cases, vendors like Amazon didn't even require a shipping label. You just put the product in a box, left it on your front porch, and the Return Fairy whisked it away, instantly adding the funds back to your credit card.

And boy, do we love easy returns! So much so that buyer behavior has changed. Not sure what size shirt to order? Just order three different ones and return the other two! On the fence about a big electronics purchase? Just order it. If you don't love it, send it back!

Returns have become so convenient that they now amount to about 10 percent of all retail purchases in the United States. Online orders get returned 30 percent of the time.

In the United States, this practice results in four billion returns worth about $400 billion, implying an average return of $100. In the rest of the world, there are an additional twelve billion returns, amounting to another $400 billion.

In the United States, 10 percent of all returns for the year are processed during the first week of January. UPS calls the day it handles the most packages shipped back to retailers "National Return Day." Normally this is in early January, but in 2018 it was December 19, reflecting the massive online retail surges caused by Black Friday promotions. According to the National Retail Foundation, over a recent Christmas season two-thirds of Americans returned something, and a quarter admitted that they had purchased something with the intention of returning it later. According to Optoro, a leading return logistics company, only about 20 percent of returned items are actually defective.

By far the most-returned items are clothing, with some U.S. clothing vendors running over 50 percent return rates. Zappos, the internet shoe vendor, embraces the order-several-and-send-the-rest-back-to-us behavior, encouraging people to use its easy returns. On average, Zappos gets back 35 percent of the shoes it sends out. But there are some customers who return *half* of the pairs they order. And, surprisingly, they are Zappos's absolute favorite customers. According to Craig Adkins, VP of services and operations at Zappos, "Our best customers have the highest return rates, but they are also the ones that spend the most money with us and are our most profitable customers." Other vendors such as Amazon (which owns Zappos) and Best Buy don't share this mindset and ban frequent returners from purchasing or from returning purchased items. There's an entire industry dedicated to helping retailers identify and ban those who abuse generous return policies.

What happens to all those returned goods? They are handled by a seldom-seen underbelly of the retail economy called "reverse logistics." This is the reversal of the normal flow of goods from manufacturers to retailers to consumers. Getting your arms around this industry is a challenge for a few reasons. First, no two companies handle their returned merchandise the same way. Second, how any given company handles returns is constantly changing. Third, how returned merchandise is handled can even vary at the individual store level. And fourth, companies are often tight-lipped about this rather opaque part of their business, perhaps because customers prefer to think the goods they're buying are brand-new. Disposing of returns has until recently been largely a nuisance, with returned goods stacking up in warehouses until the space is needed and the goods are disposed of.

In addition to returned merchandise, another component of reverse logistics is unsold goods. In addition to the 10 percent of retail sales that returns amount to in the United States, there's another 5 percent that wasn't sold in the first place. This includes fashion no longer in style, electronics that are no longer current (as we saw in the chapter on planned obsolescence), and so forth. The value of this 15 percent of U.S. retail sales that flows backward through the supply chain is roughly equal to the GDP of Poland.

Although many companies handle these returns differently, most follow a four-step triage process. Let's look at each step.

When a return comes back, the first question asked is "Can this be sold as new?" If so, it goes back on the literal or virtual shelf. But less than half of returned goods make this cut and later sell at full price. In addition, many items are opened, used, or have damaged boxes. The question of whether or not something is "new" is complex. It's illegal to sell used goods as new, but there's also no widespread definition of what "used" is.

But what of the goods that can't be sold as new? They are pushed along to the second step: to be sold as used, damaged, or returned. One example of this is Amazon Warehouse Deals, which the company describes as a "trusted destination for customers to find deals on used products at a discounted price." Other companies sell their returned goods in-store, marked as returns or open boxes. Some vendors have dedicated outlet stores designed to sell their returned goods. This is part of Best Buy's strategy. The company has thirteen stores that solely handle returned merchandise. Purchases from these outlets are not returnable unless they are defective.

In addition, some retailers have processes by which they refurbish and sell returned merchandise. Often these goods carry the same warranties and return privileges as new merchandise. Anecdotally, more than a few computer purchasers seek out refurbished units, maintaining that the testing processes are so rigorous that the items are statistically more reliable than new ones.

Unsold inventory is often sold through these channels as well. In the past, clothing retailers would sell this inventory down a step in the fashion pecking order to what are known as off-price companies. These include Ross and TJ Maxx, retailers who sell well-known brands at bargain prices, usually averaging around $10. Today, however, retailers are more often opening their own off-price companies. Nordstrom has Nordstrom Rack, for instance, as a secondary market to offload its unsold merchandise. Saks has its Off 5th stores, Macy's has Backstage, Bloomingdales has Bloomingdale's Outlet, and there are many more. These outlets are seeing strong growth. Nordstrom Rack already has more stores than Nordstrom and continues to open new locations.

Goods (both returns and overstocks) that can't be economically sold through any of these methods are sent to the third step, where they are bundled up and sold in bulk. This is the most interesting and fastest-growing method of disposal.

There are several online liquidators that sell returned merchandise and overstocks by the pallet. These include B-Stock, BULQ, and Liquidation.com. While they technically sell business to business, anyone can go to these sites and bid in an auction format on pallets of B-Stock goods. Some eschew the auction format and simply list the pallets at heavily discounted prices. You can specify a category, such as housewares, clothing, electronics, and so forth, and sometimes a condition, be it returns or overstock. You can see a photo and often a listing of the items along with the aggregate retail value of the lot. But that's about it. You don't know if all the parts are there, if the items work, or anything else. Needless to say, returns are not allowed, but discounts are steep—80 percent off is not uncommon. Browsing these sites is a little addictive, and the internet is full of fun videos of people purchasing lots off these sites and unboxing them at home.

What happens to these goods? Well, they go in a million directions. Many people buy them and then laboriously list the items individually on eBay or Amazon Marketplace. Others test and potentially fix the items to sell them on other platforms. Some lots are shipped overseas, where the products are sold in less affluent markets.

The so-called secondary market—that is, the post-retail channels that returned and unsold goods are routed through—is vast and growing. Dr. Zac Rogers of Colorado State University studies it as much as anyone. His team of researchers divides it into categories, including pawn shops, outlet stores, salvage dealers, online auction houses, charities, flea markets, and off-price vendors including dollar stores. He determined that back in 2008 this secondary market totaled $310 billion in the United States. But just eight years later it hit $554 billion, up about 80 percent over just eight years. He points out that this is 3 percent of the United States' GDP.

The secondary market does compete with the primary market. Someone who buys a scratched and dented waffle maker off eBay doesn't buy one at Walmart. Consumers now have so many channels to purchase from that traditional retailers—the ones with famous names that have been in business since Grover Cleveland's second term—are going out of business. As these retailers close more stores, they pile more goods into the secondary market. What is the value proposition of a Sears or JCPenney in a world with Amazon, eBay, Walmart, Amazon Warehouse Deals, Nordstrom Rack, and countless discount stores? Traditional retailers are desperately trying to answer this question. JCPenney, as an example, announced in late 2019 that it had partnered with thredUP, an online consignment store that sells "like-new styles from leading designers and brands." In other words, JCPenney will begin to sell used items alongside new ones. Of course, this strategy couldn't compete with the march of progress—or with the events of the pandemic—and Penney declared bankruptcy in July 2020.

Goods that can't be sold in either the primary or secondary markets end up at our fourth stage, the landfill. Optoro, the return logistics company mentioned earlier, estimates that in the United States about 5 billion pounds of returned goods end up in landfills. This might account for as much as a quarter of all returns. Why? Sometimes goods are simply too damaged to be returned, or they are in categories that can't be resold after they are returned—underwear, say, or perishables. Often these goods aren't worth the price to ship them from the retailer back to the manufacturer. This is an incredibly frequent occurrence in brick-and-mortar stores, where the items are just put in their dumpsters. Relatively few people do it, but a comfortable living can be made just retrieving things from dumpsters behind retail stores and selling them (although it should be pointed out that dumpster diving is illegal in a few places). A 2015 *Wired* piece by Randall Sullivan profiled one such person, Austin's Matt Malone, a successful computer security consultant who dumpster dives part-time as a hobby.

We chatted with Malone about his unorthodox side gig. By his own reckoning, Malone could make around $75 an hour indefinitely with a simple business model: Dumpster dive for a day or two, list it all on Amazon, then package everything up and send it to them to warehouse and distribute. He says it really is that easy. He's often tempted to do more of it, but he loves his day job too much. Even in the time he can devote to this sideline, he is constantly—and we emphasize the word "constantly"—finding amazing things in the trash. He told us the story of deciding he wanted a paddleboard. So he checked the dumpster of a local paddleboard company until he found a $1,300 one it had thrown out.

His thoughts on dumpster diving are divided in a left-brain versus right-brain manner. On the one hand, he gets an endorphin rush from finding a cool item and then selling it for a sweet profit. But the other side of his brain is always telling him there's something wrong with this picture. He thinks there's a fundamental problem with society and that "every civilization that hits the level of gluttony and waste that ours does has fallen apart. Our society is falling apart. Do we want to be like Easter Island and one day wake up to find we have cut down all the trees?"

It's one thing to return an item because it didn't fit, or because it was the wrong color, or simply because you changed your mind. But some use the return system for other reasons. According to the National Retail Foundation, about 5 percent of all returns are fraudulent. The biggest component is the return of shoplifted merchandise, which seems doubly cruel. Another is employee fraud. One of the authors vividly remembers as a teenager being a stocker for a major retailer and having to sit through extensive training. At one point, they brought in the security guy, who told all the staffers, "Now, I don't want any of you to pull the knob off a TV, then go to your manager and ask if you can buy the TV without a knob at a discount." He continued to describe all the ways he didn't want employees to defraud the company through returns, including returning merchandise pulled from the dumpsters out back. Frankly, none of his methods had occurred to any of us until he explained them, and then it was easy to think, "I could totally get away with that." It was an academic exercise, but it drove home the point that it's easy for employees to steal via returns. Another major source of fraud in returns is known as wardrobing—buying clothes, wearing them, returning them, and repeating the process indefinitely. While this is not technically illegal, retailers still consider it fraud. And it's quite widespread. By one estimation, a third of all fraudulent returns are from wardrobing.

Liberal return policies are a system built on trust. The merchant trusts that the customer isn't buying that TV *just* for a Super Bowl party, or that dress *just* for a dance at a wedding. If there was widespread abuse, then the policies would no longer make financial sense for the companies to offer. The good news is that return fraud, while measured in the billions of dollars, is still a tiny part of the whole pie. But there are the usual, and unusual, exceptions. In 2019, there was a case where a twenty-two-year-old man in Spain scammed Amazon out of $370,000 by ordering items, weighing them, removing the item, putting the same weight of dirt into the box, then shipping it back for a return. It was only after some time that an Amazon employee opened one of the boxes and discovered the ruse. But you can imagine the people who bought a pallet of returns in the secondary market only to get boxes filled with dirt. Again, the infrequency of these sorts of scams allows the system to stay afloat for the rest of us.

What about the environmental impact of all these returns? It's no doubt significant, since we're talking about fifty million returns every day worldwide. There's the truck that picks up the return, or the car to drive it back to the retail store. There's the packaging used and the label that's printed. Once the return gets back where it is going, it has to be shipped again, whether to the secondary market or elsewhere. From there, it likely has to be shipped again. And so on. But is it waste? Technically, no. Though it would be nice to figure out another way to handle it all.

Other methods of handling reverse logistics are being explored. One idea is to get people to ship the returned item to the next person who wants to buy it, instead of shipping it back to the retailer. Another is for the secondary market to actually receive the returns itself and sell them. There are many inefficiencies in the system we presently have, and given what a comparatively large part of the economy it is, this means there's much profit to be gained from increased efficiency.

Food Waste

Certain types of waste evoke more emotional reactions than others. Food waste, for instance, hits very close to home. Parents urge their children to clean their plates because there are starving children (somewhere); the geography may vary, but famine seems to always exist in the world. The idea that vast amounts of food get wasted in a world where a person dies of starvation every four seconds seems at least a tragedy and at worst downright evil.

But how much food really gets wasted? And why? Does food waste contribute to the problem of worldwide hunger, or does the phenomenon have other causes? What can be done about food waste? No broad consensus exists on any of these questions. But by looking at the mechanics and implications of food waste, we can draw our own conclusions.

To begin with, how much food is wasted? That would seem like a knowable thing. But it isn't, for two reasons. First, no one is keeping an accurate tally. Most numbers on the subject are either inferred or acquired anecdotally. As a result, the numbers in this chapter are largely estimates (in a way that figures regarding worldwide production of aluminum aren't). However, we can have some degree of confidence that they're accurate enough to allow readers to draw logical conclusions about the problem.

But the second, larger reason is that we need to settle on a definition of food waste. Depending on which definition we use, the numbers are vastly different—perhaps by as much as a factor of three.

Think about it. What constitutes wasted food for you? The most conservative definition, the one that yields the lowest figures for waste, is any food that was consumable but that didn't get consumed and instead ended up in a landfill. By this tally, food waste in America comes to approximately 350 pounds per year per capita. That's just under 1 pound a day per person. And given that the average American generates approximately 5 pounds a day of municipal waste, it stands to reason that 20 percent of everything that is landfilled by weight is food. Using this definition makes a lot of sense: Landfilled food is a copious producer of methane, a potent greenhouse gas. So food that makes its way to dumps is a particularly insidious form of waste, since not only is it not consumed, but it actually winds up doing harm as well.

On the other end of the spectrum is a far more expansive definition of food waste. In this estimation, food waste consists of anything removed from the supply chain that is *or ever was* fit for human consumption.

This broader definition includes food grown as feed for animals, food used to make biofuel, food never harvested and plowed under for fertilizer, food destroyed by inclement weather and pests, and much more. If at any point a human could eat it, but doesn't, then a plausible argument can be made that it was wasted. There are good reasons to use this definition as well, especially if you believe that redirecting corn to make biofuel is inherently bad in a world where people go hungry. When people use this more expansive definition, it's fairly obvious by their choice of words. They say things such as "Half of all food grown never reaches a human stomach." Using this definition, America's waste is as much as 900 pounds per year per capita, and other prosperous countries, while coming in with slightly lower numbers, are in the same ballpark.

Between these methods there are divergent calculations of how much food is truly wasted. To further complicate matters, some researchers measure the amount of food waste by pure weight, others use the monetary value of the food, and still others use the caloric content of the wasted food. Under one calculation, a pound of rib eye steaks that rots yields a much larger amount of waste than a pound of wheat that is plowed under. Depending on the method used, estimates of how much food truly gets wasted range from anywhere between 15 percent and 50 percent.

Regardless of which definition of waste we use or how exactly it's measured, the 60 million tons of food in the United States that go into landfills each year is a good place to begin our examination. Why does so much go in?

Sixty million tons a year sounds like a lot, and it is. It would take a farm the size of Virginia to grow that much food. Sixty million tons a year works out to the weight of the Great Pyramid of Giza every month. Of course, spread across the whole U.S. population, we get the 1 pound per person per day figure mentioned previously. That works out to look something like an apple and an avocado. But when you look at the whole, it's a staggering amount.

Individual food items that get wasted dwarf the imagination. An aircraft carrier is a thousand feet long and has a crew of 6,000, and the weight of it and the crew is roughly equal to the amount of turkey we in the United States throw out the day after Thanksgiving. Then we throw out a second aircraft carrier's weight worth of Thanksgiving vegetables as well. It doesn't look like all that much in *your* trash can, but the problem is that everyone is doing it.

Yet, culturally, wasting food is something people go to great pains to avoid. The origin stories of hundreds of beloved dishes today, from fried rice to bread pudding to corned beef hash, all begin with a question of what to do with some leftover food item that would otherwise be discarded.

Back in the 1950s, at a Mexican restaurant in Disneyland, a tortilla sales rep noticed a bunch of stale tortillas in the trash. He mentioned to the management that they should cut them and fry them up. They did so, seasoned them, and Doritos were born. In 1953 Swanson massively over-ordered turkey for the Thanksgiving season and was left with half a million extra pounds that it didn't know what to do with. Luckily, a salesman named Gerry Thomas suggested they make a version of the three-tray plate used for airplane food and fill it with a frozen dinner to sell to the public. Thus, the turkey was put to good use—and the TV dinner was born. Industrious butchers who manage to put all of a pig to use are said to "use every part except the squeal." So it isn't that we're collectively blasé about food waste; we go to great effort to avoid it.

It would be nice if there was one simple, easily fixed reason so much food gets landfilled, but there simply isn't. If we had to pick a single root cause that explains the volume of food waste while simultaneously explaining why Americans throw out more food than people from other countries, it's that food has become remarkably inexpensive. Over the last half century, food prices in the United States have fallen by half. And during that same period, the amount of wasted food per capita has grown by 50 percent. From an economic perspective, the value of the food that is landfilled is roughly $175 billion. That's a big number, but if we spread it out over a population of 350 million, it's $500 a year, or

about $1.25 a day. People in the United States simply aren't will-
ing to change their behavior very much for a buck and a quarter
a day.

How is it that food is so cheap? For ten thousand years it took
nearly 90 percent of all humans to grow enough food to feed humanity. Now, in the prosperous West, it takes about 3 percent.
Automation is most of the answer, of course. Also contributing
are genetically modified or selectively bred crops, which account
for much of what we grow. They are not a new phenomenon;
farmers knew for thousands of years that cross-breeding particularly productive plants would result in more generous offspring.
And for as long as we have known how to make X-rays, we've
been shooting them at seeds hoping something magic happens,
and every now and then it does.

There are over fifty thousand edible plants, but two-thirds of
our calories come from just three: corn, rice, and wheat. Modern
food prices are kept low also because our crops are so specialized
for production. An old adage holds that "nothing grows in a
cornfield except corn," and while that's not strictly true, we are
quite efficient at growing the things we eat. A century ago, a
farmer might have grown thirty bushels of corn on an acre; today
we grow five times that much. But as efficient as the United States
is, it pales next to the number-two food exporter in the world
based on value.

Care to guess who that is? The Netherlands, that densely populated country scarcely bigger than Maryland. How do they do
it? Through even more incredible efficiency. They grow food in
dense, climate-controlled greenhouses, using farming methods
that require just 3 percent of the chemicals used in the United
States. They double yields of many vegetables by precision
farming—that is, individually caring for each plant according to
its needs. In so doing, the Dutch are the world's largest exporter
of onions and potatoes. These innovations, and hundreds more,

give a glimpse into how we can ultimately continue to drive down waste in the food production process.

Other factors keep food inexpensive. When child labor laws were passed in the United States, specific exceptions were included for the harvesting of crops. While you have to be sixteen to get a job filing papers in an air-conditioned office, a twelve-year-old with parental permission (or a fourteen-year-old without it) can pick food out in the sun. Under some circumstances, even the twelve-year-old minimum doesn't apply; anecdotally, children as young as six have been spotted in the fields in the United States. Agriculture workers are often exempt from minimum wage laws as well. The exceptions in the 1938 law were well-meaning, making sure that in a country of family farms parents weren't breaking the law when they got their kids to help out with chores. However, these laws remain on the books, and a quarter of the produce grown in the United States is picked by about half a million kids, helping to keep prices low. The agricultural sector in America is also reliant on the low-cost manual labor of adults, many of whom are undocumented immigrants, working for low wages without many of the protections afforded those who work legally.

Finally, agriculture is heavily subsidized in the United States, to the tune of about $20 billion a year. Some argue for the practice on national security grounds, pointing out that a nation that can't feed itself can't defend itself, either. These subsidies also drive prices down.

If one of our definitions of waste is edible food that ultimately ends up in a landfill, it's important to note that it's not just the leftovers from dinner at home that wind up as refuse. There are four primary sources of wasted food: processors, restaurants, grocery stores, and consumers. Let's look at each to see where the waste truly comes from.

First, processing and production. There are several ways food loss happens at this stage. Some amount of food is culled and discarded due to appearance, shape, or size. This can be a substantial amount for certain fruits and vegetables. In some industries, there are secondary markets to sell this food into, and entrepreneurs are creating more. If every pork chop was perfect, goes the saying, there wouldn't be any hot dogs. But in many cases, aesthetically challenged produce goes to waste.

Ever on the lookout to reduce waste, savvy producers have figured out ways to repurpose their offerings to take advantage of imperfect produce. Misshapen carrots are now routinely peeled, cut, and sold as "baby" carrots, a pretty cruel-sounding name if you stop and think about it.

But even with those baby carrots, food is lost when vegetables or meat are trimmed, peeled, or butchered. When fruit is turned into juice, there's waste, as well as when milk is pasteurized, foods are canned, or products are baked.

Similarly, there is loss during distribution. In late 2019, a ship carrying sheep from Romania capsized, causing nearly all of the fifteen thousand animals on board to drown. Distribution waste happens for a number of reasons absent tragedy, however, from poor logistical planning to changes in demand at the store level. And every load of perishables is, in effect, a time bomb ticking its way toward its expiration date.

Next come restaurants. As consumption patterns change, restaurants become a larger and larger source of food waste. Not long before the onset of the COVID epidemic, for the first time in history, Americans spent more at restaurants than at grocery stores. How do restaurants waste food? Food is often lost in the kitchen prior to cooking—dropped or burned. Then some amount of food is cooked but not sold. Restaurants serve ever-larger portions to consumers to promote the "wow" factor when the food is brought out, and to make for nice Instagram shots.

Even with the ubiquitous doggie bag, restaurant patrons on average leave about one-fifth of their food uneaten, which means many will hit their $1.25 daily waste quota right there. With all these factors taken together, for each meal eaten at a restaurant there is half a pound of food waste, and despite efforts by some restaurants and charities to get this food to the needy, about 90 percent of all wasted restaurant food goes into landfills.

The COVID pandemic had a two-part effect on food waste. Early on, it skyrocketed. The food supply chain was engineered to sell half its food in grocery stores and the other half to restaurants, cafeterias, stadiums, cruise ships, and other commercial enterprises. When that second half plummeted, much of the food heading to those places was wasted. The Dairy Farmers of America estimate that in the United States, half a million gallons of milk a day were destroyed. One single processor ended up destroying 100,000 unhatched chicken eggs—a day.

But the second-order impact of the pandemic has been a huge decline in food waste. Now that consumers are making more meals at home, they're also more likely to be eating the leftovers. While restaurant portions are supersized, serving sizes are more realistic at home. This observation is more than anecdotal. As grocery stores experienced shortages, families had incentives to manage their food consumption more efficiently. In the United Kingdom, by one estimate, the percentage of basic staples— bread, milk, potatoes—thrown out by consumers fell by half during the pandemic.

Grocery stores generate food waste as well. While writing this chapter, one of the authors had the experience of being in one of Austin's nicest grocery stores watching dozens of employees throw away every single item in the dairy, bakery, and packaged meat sections. All the milk, all the bacon, all the eggs, all the cheese, everything. Early that morning, there'd been a small fire in the back of the store, and in the aftermath food inspectors swabbed everything. Anything with smoke residue on it had to be discarded. The dumpsters out back, by the way, were locked up, so no one could purloin any of the pricey bacon. That most of the bacon was already smoked added to the irony. Dumpsters at grocery stores are often required by law to be inaccessible to the public due to the health risks of eating, say, day-old raw shrimp.

Of course, fires in grocery stores aren't an everyday occurrence. But the fact they can happen—and the reaction when they do—underscores how particular grocery stores are about the food they sell. Despite a worldwide "ugly food" movement, many stores remove and discard produce that doesn't look perfect. They also throw out food nearing its "best by" date, even though that date has little bearing on whether the food may be safely consumed. They throw out most prepared food each night before closing. In addition, they throw out food that doesn't make it off the shelves in time to clear the way for newer goods.

Where does home food waste come from? A lot of it, quite frankly, comes from imperfect information that sellers have little incentive to make clearer. One of the primary drivers of confusion is the various dates used on packages. Some packages have a "best by" date. Others say "sell by." Some say "use by." What about "packaged on" or "enjoy by" or "better before" or "guaranteed fresh until" or "discard on"? Sometimes there is simply a date printed on the packaging with no explanation at all. Consumers, not really sure what most of these mean, conservatively assume that the implication is "will make you sick after." Thus, many adopt a "when in doubt, throw it out" attitude. The dates also have the subtle effect of assuaging guilt over wasted food. Yes, you intended to cook something with it, but the date has passed and now your hands are tied. It's not your fault.

But it isn't just the "best by" dates that cause problems. At the grocery store, consumers often get ambitious about what they will cook in a given week and buy the makings for gourmet feasts.

But then life sets in, and by the end of the week they're ordering a pizza. Some of the ambitious food can be frozen or otherwise preserved, but much is thrown out. When food was dearer, that unused beef might have been salted, those cucumbers pickled, or those berries turned into jam for later consumption. But when food is so cheap and plentiful, why bother?

When consumers do cook, they invariably cook more than they need. Again, why not? Food is so cheap. Then there are inevitably leftovers, which are packaged up with the best of intentions—and promptly forgotten. By the time they're seen again in the back of the refrigerator, their only potential use is for a high school science fair project.

Interestingly, the healthier your diet, the more food you likely waste. Healthy diets include lots of fresh fruits and vegetables, over *half* of which gets thrown out since they spoil so quickly. The only other category in which more than half is wasted is seafood, which often doesn't pass the literal sniff test.

The closer to the consumer level that waste happens, the more costly and impactful it is. If a farmer fails to pick an ugly carrot or picks it but throws it away immediately, little is lost. But a food item that successfully runs the gantlet from field to processor to grocery store to your home, and then gets thrown out after a dinner guest leaves it on their plate, isn't just wasted food. It's wasted energy, wasted labor, and wasted time, even if you compost it and use it on your roses.

Are there easy fixes to any of this? Technology has certainly helped in the past. The ability to refrigerate food might even qualify as the biggest advance against waste in human history. Likewise, there are coatings to spray on fresh vegetables that could double their life. Cheap sensors will eventually be able to tell if a given food item is safe, regardless of its "sell by" date. Growing or 3D-printing meat will reduce the amount of waste from mistakes made in butchering animals. To minimize the impact of waste, some advocate that the planet should switch to flour made from ground-up insects, or, perhaps more palatably, vegetable-based meat substitutes. Even switching among different kinds of meat has the ability to lessen the ecological impact of food production and therefore of wasted food; on a per-calorie basis, for instance, poultry requires significantly less resources than beef. And many vegetarians and vegans adopt those diets not simply for perceived health benefits but for environmental reasons, pointing out that huge amounts of food are grown simply to be fed to animals to produce vastly smaller amounts of food.

So how should we look at the $500 of wasted food per person per year? Is that just the price we pay for having safe and nutritious (or perfectly shaped) food? Some of it is clearly a choice; can that truly be thought of as waste? If a dozen ugly carrots have to be thrown out so you can enjoy a perfect one, and you're willing to pay for those dozen rejects in the form of higher prices for the beautiful one, the definition is subjective. We wouldn't apply that same standard elsewhere. If half of the pots a pottery factory made were aesthetically flawed, though usable, and were subsequently thrown out, we wouldn't wring our hands in anguish over the "wasted" pots. We would simply see it as the price of doing business.

Somehow food is just . . . different.

As mentioned before, wasted food has a uniquely moral aspect to it. When we were kids and our parents told us that we should eat our food because children were starving elsewhere in the world, that admonition came from an empathetic place in the human heart. And although linking what's on your dinner plate with hungry kids in another country is so logically flawed that it even has a name (the fallacy of relative privation), it's a deeply human reaction. A computer would never fall for the kind of false logic our elders inflicted on us when we were kids, and this just underscores how different (and, we would argue, better) humans are than machines. It's this sentiment, as logically flawed and entirely human as it is, that explains why the French Senate unanimously passed a law banning grocery stores from throwing away food but did not pass a similar law regarding pottery.

There are about 800 million hungry people in the world. Why? It's not because the world has trouble producing enough food for its population. We produce many more calories than we need to feed everyone. And it's certainly not because you didn't finish your green beans last night. The sad truth is that nearly 80 percent of all hungry people in the world live in countries that are net food exporters. It turns out that in the modern age, you starve to death not because you have no food but because you have no money. In the end, our biggest problem isn't that we throw away food but that we choose ways to allocate it that leave many people hungry. Our challenge isn't agriculture but ethics.

Yet morally demonizing all food waste or attempting to ban it by fiat might lead to even more waste. As we've seen time and time again, there are unexpected consequences to even well-meaning attempts to reduce waste. It may well be the case that to *not* waste any food would end up costing more than $500 per person. You could drive to the grocery store every day instead of once a week, and in doing so waste fewer vegetables. But how much gas would you waste instead? Presumably if money could be saved by not wasting food, the agriculture industry would do it. If potato skin shavings at the potato chip factory could be swept up and sold as pig fodder profitably, it stands to reason that the factory would do so. But if the energy needed to get those shavings to the pigs exceeds the energy needed to grow other food for the pigs, what have we really gained? It may be worth it to a farmer to plant 20 percent more of a crop than they can sell as insurance against the possibility of a bad harvest, knowing full well they may just plow that extra crop under.

Food waste, in some ways, is a testament to just how productive the earth is. Our world produces so much bounty that it doesn't even make financial sense to keep from wasting some of it.

So that's the irony of our situation. We make more food than we can eat. It costs ever less. Millions around the world starve while we throw out massive amounts of good food. Obesity now kills more than hunger. Our food choices are making some healthier than they've ever been, while the choices of others are killing them.

The world's population is expected to peak at 11 billion, about 50 percent higher than it is right now. There's no question that with existing techniques we can produce enough food to feed that many people, but to make sure that 11 billion people actually get fed will require societal changes.

And yet, the challenge of production is not trivial. A report by the World Wildlife Foundation says it best: "Humanity must now produce more food in the next four decades than we have in the last 8,000 years of agriculture combined. And we must do so sustainably."

The Distance Dilemma

I f you were to ask one hundred ecologically conscious people
what their friends and neighbors should be doing to reduce
waste in their lives, the first answer for many would be to suggest
buying locally grown produce, rather than food trucked in from
hundreds or even thousands of miles away.

The notion of eating food produced close to home has become
much more popular over the past decade—so much so that a
word, "locavore," was coined to describe those who subscribe to
this philosophy. It was deemed the U.S. word of the year in 2007
by Oxford University Press. ("Carbon footprint" edged out "locavore" to take the title that year in the United Kingdom, perhaps
aptly, since the notion of reducing one's carbon footprint is often
integral to the decision to eat locally produced foods.) Grocery
stores now mark products that are produced nearby, and many
restaurants tout themselves as using organic, sustainable, locally
sourced products.

But are local products truly less wasteful? At first glance, the
notion seems obvious; buying locally grown peas in Los Angeles

must be less wasteful than buying peas from Peoria—or worse, Perth. But our intuition is notoriously deceptive when it comes to evaluating the truth behind such claims. If we break down the energy costs of moving food from a faraway location to growing it closer to home, we can get a clearer picture of whether eating locally truly is less wasteful.

The way the word "calorie" is used in the context of science (the amount of thermal energy required to raise 1 gram of water 1 degree Celsius at standard pressure) is a bit different from how it's used in the context of food. Food calories are actually kilocalories—one thousand times larger than science calories. Thanks a lot, science namers.

In this chapter, you'll see the capital-C version of Calorie when we're talking about food calories. By using Calories to keep score, whether we're talking about energy that comes from food or from another source, we can compare the "cost" of moving our food to growing it closer to home.

Moving that food most likely uses some kind of fossil fuel. And like the fat in your body, gasoline is a store of energy. It's relatively easy to figure out how many Calories are in gas; when it's burned, we can measure how much heat it can add to water.

It turns out that fat and gasoline have almost the same energy density. Where 1 gram of fat in your body stores 9 Calories, 1 gram of gasoline has 10. A full gallon of gasoline weighs in at about 6 pounds, and therefore has the equivalent of about 30,000 Calories.

If your car gets 30 mpg, it means it "costs" 1,000 Calories of gasoline to go a mile. A person, on the other hand, can bike that same mile for just 50 Calories and walk it for 100. It might sound like your body is more efficient than a car, but remember that the car weighs 4,000 pounds, can move at highway speeds, is air-conditioned, can play music while you drive, and allows you to bring friends with you for only marginally more energy. However, your body is also powering a supercomputer—your brain—the likes of which your car can never even aspire to.

When you break down the Calories required to move 1 pound 1 mile, you get close to parity. Humans and cars are about equally energy efficient. Compared to food, however, gasoline is a remarkably inexpensive source of energy. If a human could drink a gallon of gas and get those 30,000 Calories directly, it would run their body for two weeks at a financial cost of just 20¢ a day. (Please, don't do this.)

Across the entire U.S. food ecosystem, for each Calorie a person eats, about 10 Calories are required to grow, transport, and store that food. That means that the 3,000-Calorie pizza you had delivered last night ultimately took 30,000 Calories to make its way to you. That's about a gallon of gas. The energy expended to create and then bring all the ingredients together, cook that pizza, then deliver that pizza to your door is ultimately ten times the energy in the food item itself. A person eats, in round numbers, just under 1 million Calories a year. Multiply that times the entire population of the United States and you get about 300 trillion Calories of food needed to feed us all annually. For every Calorie of food that we eat, there's an additional ten Calories of energy expended in various fuels.

A 2008 Cornell study found that trucking in vegetables from the Midwest to New York uses one-sixth of the energy it takes to grow them in greenhouses on the spot. Why? Greenhouses take lots of energy, and great economies of scale can be had in the vast lands of the West. In the overall scheme of things, in this instance, trucking produce halfway across the country uses less energy.

This is true even when the two endpoints are half a world apart. One study conducted by New Zealand academics compared the environmental impact of someone in the United Kingdom consuming foods produced there versus consuming items imported from New Zealand. They found that "the energy used in producing lamb in the UK is four times higher than the energy used by NZ lamb producers, even after including the energy used in transporting NZ lamb to the UK," largely because of the extra costs of energy and fertilizer needed to raise lamb in the United Kingdom.

Or consider the case of banana production in Iceland. Incredibly, there is such a thing—or rather, there was. Iceland, as you may know, is far away from most of the world's banana-growing regions. Really far. But Iceland is bubbling over with geothermal energy, which is relatively inexpensive compared to other energy sources elsewhere in the world, as mentioned in the chapter on aluminum smelting. Given its remote location and its inexpensive energy, you might think that perhaps these two facts would make for inexpensive, efficiently grown Icelandic greenhouse bananas.

Alas, such is not the case. They tried it once, and it was a disaster. Once government tariffs on imported bananas were removed, the indigenous banana industry in Iceland collapsed. It turns out that no amount of cheap energy or distance from banana sources can overcome the fact that Iceland borders the Arctic Circle and therefore gets very little sunlight. Bananas in Iceland take two years to mature, unlike their equatorial cousins, which are ready to consume in mere months. And no matter how inexpensive geothermal energy is, in major banana-exporting countries such as the Philippines, Costa Rica, and Ecuador, sunlight is, well, free. Thus, buying locally grown bananas in Iceland doesn't save on overall energy costs.

Arctic bananas are an extreme example, but even local farmers' markets don't automatically make the energy efficiency cut. After all, each individual farmer has to drive their produce in from all around the area. Then the customers have to drive in as well. However, there are significant economies of scale when it comes to moving food in bulk. As Australian academics Els Wynen and David Vanzetti point out in their paper on the limitations of using miles traveled as a measure of efficiency in the food system, transporting 10 tons of food 1,000 kilometers in a single 10-ton truck uses less energy than the same amount of food traveling the same distance in twenty half-ton trucks.

Not only does each truck burn a lot of fuel getting to the market, but customers still have to shop at their regular grocery stores as well to get other staples. So trips to the farmers' market are *in addition* to the trips to the supermarket, not a replacement for them. Every trip to the farmers' market wastes fossil fuels because equivalent items can (or could) be found in highly efficient centralized supermarkets.

Of course, just because something is on its face wasteful isn't necessarily a reason not to do it. Supporting local farmers is of great value to many who want to see family farms prosper and thrive, and who want to support sustainable practices. There are surely more than a few people in Iceland who would pay an extra nickel to walk around Reykjavik eating a locally grown banana because, well, it's kind of cool to eat a banana grown in Iceland by geothermal energy over the course of two years.

But if much of the value of buying local or frequenting farmers' markets consists of intangibles, such as supporting local farmers, those intangibles deserve at least some passing scrutiny. Are farmers' markets what they seem to be? Is the image of the Norman Rockwell–esque family-run microfarm selling their produce directly to their neighbor at the Saturday market really accurate?

Journalist Laura Reiley wrote a scathing series of exposé articles called "Farm to Fable" for the *Tampa Bay Times,* for which she became a Pulitzer Prize finalist. Reiley would go to farmers' markets, ask the vendors if they grew their own food, then actually drive out to where they said their farm was. Sometimes things were exactly as described, but often they weren't. She documented a wide range of practices that were at the least misleading, but mostly out-and-out lies. Often the vegetables were grocery store rejects with the labels peeled off or were simply purchased from outside the country and "rebranded" as locally grown. "Organic" strawberries might have changed hands a couple of times before arriving at the market, with their provenance murky at best. Restaurants were no better: They were offering "locally grown" foods that weren't actually local. Fish sold as one species was really another, pricey water bottles were refilled from the tap, "house-made desserts" were store-bought, and on and on and on.

At the end of the day, the final verdict isn't that buying local is *inherently* wasteful; it's that the picture isn't nearly as simple as it seems.

It's possible to go in circles indefinitely on the trade-offs between efficiency and sustainability in a food context. The same Australian study cited earlier takes a look at what transporting organic food by air costs in terms of carbon emissions. Unsurprisingly, as you saw in our chapter on air travel, introducing airplanes into the equation basically blows anyone's carbon budget. On the other hand, the study determined that if air travel were unavailable, the ripple effects would be significant and would lead at least some producers of organic produce to revert to conventional production.

Overall, the belief that calculating the number of miles our food travels before it hits our tables is a good way of evaluating how much waste and inefficiency is involved in its production is just not true. Food transportation networks are incredibly efficient, so much so that energy used to transport food is a trivially small part of the total energy used to make our food.

As we saw in the case of the Icelandic bananas (and in this book in general), if all we look at are the first-order effects of waste—in this case, the extra fuel required for transportation of food—the picture seems clear: Grow food as close to where it is eaten as possible. However, such an analysis fails to examine the higher-order impacts of local food production, specifically with regard to how much more energy, fertilizer, water, and land is necessary to produce any given food. When the analysis is expanded, the easy answers often change.

PART 3

The Science of Waste

Deep Breath, and . . . Exhale

It would be impossible to write a book on waste that didn't cover carbon dioxide. It's in the headlines nearly every single day. To many, CO_2 emissions are the single worst kind of waste, due to their role in climate change and all that could come with it, including the acidification of the oceans, rising sea levels, mass extinctions, and planet-wide social unrest.

But oftentimes, the story of CO_2—where it came from, how it's emitted, how it can be lessened—is told in isolated pieces, leading us to know more about its effects than we do about the underlying systems that drive them.

Is the carbon dioxide that humans exhale in the respiratory cycle waste? There's a clear answer to that question—but to truly understand that answer requires a deeper understanding of CO_2. This chapter might not contain everything you want to know about carbon dioxide—but it contains most of what you need to know to understand this critical gas in the context of waste.

To tell the full story of CO_2 requires going back in time 4.5 billion years, to when the earth was cooling. Volatile elements within the planet escaped and formed the early atmosphere via a process known as outgassing. While there is a debate about the exact composition of the early atmosphere, one prominent theory suggests that it consisted of hydrogen, methane, carbon monoxide, and carbon dioxide.

Such was the state of things for several hundred million years. Then, about 3.8 billion years ago, the oceans began to form. Life on the planet existed at this time, but in a form akin to anaerobic bacteria—that is, creatures that can only survive in an environment *without* oxygen.

Ocean water has a number of gases dissolved into it as a result of continuous natural reactions that take place wherever the surface of the water meets air. Even today, carbon dioxide leaves the air and dissolves into the oceans. But billions of years ago, the atmosphere had so much more CO_2 that massive amounts entered the seas. While some of it remained a gas (producing carbonated water), most reacted with the seawater to form carbonic acid or, more commonly, bicarbonates. Life-forms used that carbon to grow.

When those life-forms died and sank to the bottom of the ocean, they became the limestone and fossil fuels we use today. The cycle continues today: While the oceans themselves give off some CO_2 into the atmosphere when animals die and decay, they absorb more than they emit, storing it away for eons to come.

When the oceans first formed, there was essentially no oxygen in the atmosphere. How we know this fact is pretty interesting. While ice cores can be pulled from glaciers and melted to get actual samples of ancient atmosphere, the oldest ice cores go back "only" 2.7 million years. However, we can infer the lack of oxygen in the atmosphere from billions of years ago because the iron we find in the most ancient rocks isn't oxidized. Oxygen is highly reactive stuff; it readily combines with many elements, and it's hard to maintain in its gaseous state in nature because it wants to react with—that is, oxidize—everything. In short, no rusty iron ore from the distant past means no atmospheric oxygen then, either.

About three billion years ago, something massive happened. A biological process appeared by which microorganisms could pull the plentiful carbon dioxide from the atmosphere and combine it with the water of the oceans and sunlight to produce energy. That process, oxygenic photosynthesis, produces an otherwise unneeded waste product that you can probably guess from its name: oxygen.

Photosynthesis was a huge competitive advantage to these early microorganisms, and they thrived. But mysteriously, for about a billion years, we had photosynthesis but still not much atmospheric oxygen. No one is exactly sure why.

About two billion years ago, however, the atmosphere started flooding with oxygen. Unfortunately for them, this oxygen proved deadly to much of the anaerobic life-forms on the planet. Some still exist, and we can find remnants of those early life-forms in today's extremophiles, which thrive in the deep ocean away from light and air, living off sulfates emitted from thermal vents.

Eventually, the level of oxygen in the atmosphere stabilized to where it is now—about 21 percent. The mechanism by which this stasis was produced remains a mystery. James Lovelock's Gaia hypothesis speculates that the earth's life, the entire biosphere, is a self-regulating complex system that keeps critical systems like oxygen concentration regulated within ranges conducive to life—in much the same way as the human body regulates its temperature around 98.6 degrees Fahrenheit. But this theory is speculative.

About half a billion years ago, algae from the ocean made its way onto land, where it became moss. From there, we got plants with roots. Then things went a bit crazy, at least from today's perspective. Had you been around five hundred million years ago, you would have seen a land covered with lichens, giant ferns, and three-story-tall mushrooms. Trees came much later, so much so that there were sharks on the planet before there were trees.

The carbon cycle that exists today is relatively straightforward: When a plant grows, it absorbs carbon dioxide. Then it dies, and through the process of decay, the CO_2 is released back into the air. But when the first land plants appeared—and for tens of millions of years later—the specialized bacteria that helps dead plants decay *didn't exist yet*. This is important. Plants would grow, absorb carbon dioxide, then die. They would fall over, never fully decay, and thus *never release their carbon* back into the atmosphere.

A new plant would then grow on top of that organic material, where eventually *it* would fall over dead and not decay, and not release its carbon, either. This process repeated ad infinitum. So the land became matted with a huge amount of dead but not decayed plants. If you've ever wondered where coal comes from, wonder no more: This incomprehensibly large heap of plant material became coal deposits. Credible theories posit that each foot of coal used to be 15 feet of matted dead plants, so a coal deposit 40 feet thick was once a 600-foot-high heap of dead plant matter—a height taller than the Washington Monument.

As a result of this process, while the oceans were busy pulling huge amounts of carbon out of the air and sequestering it away, along came the land and its non-decaying plants, which sequestered even more carbon.

So far, so good.

At this point in the story of CO_2, a unit of measure called a gigaton becomes essential. A gigaton is a billion tons. A gigaton is a weight so large it's difficult to find a good reference point, but let's try: All the people on the planet together weigh less than 1 gigaton, and all the cars and trucks in existence weigh just 2 gigatons.

Alternatively, take everything in Manhattan—all the buildings, the people, the roads, everything—and weigh it. It will total one-eighth of 1 gigaton. The entire output of the Mississippi River into the Gulf of Mexico for a day gets closer, equaling about 1 gigaton of water. Or take the biggest sports stadium you have ever been in, fill it full of water to the top, weigh that water, and then repeat—three hundred times. That's 1 gigaton.

Now let's play a little game. How many gigatons do you think the earth's atmosphere weighs? More or less than 1 gigaton? Make your best guess.

The earth's atmosphere weighs about 5.8 *million* gigatons. This is essentially an inconceivable number. But if you add up all the carbon deposits that we just discussed—all the limestone, fossil fuels, dissolved CO_2 in the oceans, plant life, all of it—you get over *120 million gigatons*.

And here's the big shocker: At one point, most of this carbon was in the atmosphere.

Now, it wasn't all in the atmosphere at the same time. Nonetheless, the sheer volume suggests that for billions of years our atmosphere was dominated by CO_2, and perhaps earth was something like how Venus is today.

We'll get to human respiration in a moment, but before we do, let's talk about how humans generally consume carbon outside our bodies. Most of the energy we generate comes from burning those ancient stores of carbon in the form of coal, oil, and natural gas. This combustion uses oxygen and releases carbon back into the air in the form of carbon dioxide.

Burning 1 pound of coal generates about 1 kWh of electricity and creates about 2.5 pounds of CO_2, depending on the quality of the coal used. (As we've seen previously, burning something requires the addition of elemental oxygen, which is why the CO_2 produced is actually *heavier* than the original fuel source—it contains not only carbon from the coal but also oxygen.) Other fuels create less CO_2, and energy sources like nuclear, solar, and wind emit none directly.

Across all fuel sources, the electrical grid in the United States produces about 1.5 pounds of CO_2 per kWh. Assuming a residential price of electricity of about a dime per kWh, you can determine the number of pounds of CO_2 your home produces by taking your average electric bill and multiplying by fifteen. A $200 monthly electric bill thus equates to 3,000 pounds of CO_2 released every month just due to the electricity used within your home.

The total amount of CO_2 produced by this human activity of burning fossil fuels is about 40 gigatons a year. That doesn't seem like a lot compared to the 5.8 million gigaton weight of the atmosphere. However, the percentage of the atmosphere that is currently CO_2 is only about ½5th of 1 percent. In fact, the total amount of atmospheric CO_2 is only 3,200 gigatons, about the same weight as Mt. Everest. So even adding relatively small amounts of CO_2 changes the total amount of CO_2 in the air by a large percentage. Since there is so little CO_2 in the air, we measure it not in percent but in parts per million, or ppm. At present, the atmosphere is about 420 ppm CO_2. So if 420 ppm of CO_2 is about 3,200 gigatons, then 1 ppm is about 8 gigatons. (Remember that fact; we will return to it.)

While today's level of CO_2 is 420 ppm, before humans started burning fossil fuels at scale (say, around the year 1750), the air was only about 280 ppm CO_2. Geologically speaking, that 270-year span isn't even the blink of an eye—it's effectively instantaneous. We've changed the carbon makeup of the atmosphere by a lot in a very, very small amount of time.

What's happening now? When we go back to the 40 gigatons a year of CO_2 being released by human activity, only half of it stays in the atmosphere. The other half is sequestered in roughly even parts into the land and oceans by natural processes.

Of the 40 gigatons we release, about 20 make it into the atmosphere. Divide the 20 gigatons that make it into the air by 8 gigatons of CO_2 (which represents 1 ppm), and you come up with the result that the CO_2 concentration of the earth's atmosphere is growing at about 2.5 ppm per year. And that's exactly what's happening. Every four or five months, the CO_2 concentrations in the earth's atmosphere tick up another ppm. We can actually watch the composition of our atmosphere measurably change in time periods measured by months.

As was just noted, humans emit 40 gigatons of CO_2 each year, but only half of that stays in the atmosphere. Why is that? Here is where the story takes an interesting turn. The earth naturally emits *and absorbs* an immense amount of CO_2. The earth operates as if it were a giant organism that rhythmically breathes in and exhales CO_2. This exchange occurs in both the oceans and on land, where plants take in CO_2, die, and release that carbon, or where plants become food for animals that eventually die and give back the carbon. A few hundred years ago, before human activity began to release large amounts of CO_2, the amounts of CO_2 the earth absorbed and emitted were essentially identical, about 750 gigatons a year. This kept the amount of atmospheric carbon essentially steady for a long time. In fact, over the last million years or so, levels of atmospheric carbon have fluctuated between 180 ppm and 280 ppm. If you graph that fluctuation over time, you would see that it oscillates between those two numbers on a regular basis, and it's theorized that it changes based on the 26,000-year cycle of wobbles in the tilt of the earth's axis. Tiny changes in that tilt, combined with the earth's location relative to the sun, change the amount of sunlight different areas of the planet get, triggering periodic ice ages and warmings.

But a few hundred years ago, something changed. Humans began burning fossil fuels, lots of them, releasing additional CO_2 into the atmosphere. The 40 gigatons that we currently emit each year doesn't sound like much compared to the 750 gigatons that the earth's natural processes emit, but it's enough to throw the system out of balance.

At this point, astute readers are probably wondering the following: If the planet's carbon system is generally in balance without human activity, why don't the 40 gigatons humans emit each year stay in the atmosphere—or, alternatively, why aren't the 40 gigatons reabsorbed the way naturally produced CO_2 is? Why are 20 absorbed, while 20 remain in the atmosphere? The answer is that the earth's carbon cycle is dynamic, and it does respond to human activity. As more carbon is released by humans, more is absorbed by the earth. For instance, there are studies that suggest the extra CO_2 from human emissions is "greening" the planet, causing more plants to grow. More atmospheric carbon means more photosynthesis, which means more plants. However—and this is a big "however"—the earth's systems can't react quickly enough to offset the sudden increase human activity has caused. Perhaps they might, if we're willing to wait another few millennia. But in the meantime, they remain out of balance. The single fact that everyone agrees on is that net global average atmospheric CO_2 levels are rising, and the question of the hour is what impact that will ultimately have on the life on this planet. Little comfort can be taken in the fact that the human contribution to the carbon cycle is small relative to the whole. After all, the normal body temperature of a human is about 100 degrees Fahrenheit. Raise it to just 110, and that human dies.

At long last, we get to people themselves. What effect does human breathing have on atmospheric CO_2? Each person exhales about 2 pounds of CO_2 a day. That works out to 1 gram, about the weight of a paper clip, every twenty breaths. Multiply that out across the population of the world and you get a total of 3 gigatons of CO_2 exhaled by humans per year. As mentioned in the introduction, this process is crucial to how people lose weight.

If you lose a pound of fat, it's not that you converted it into energy and it vanished. Instead, to lose a pound of fat, you inhale 3 pounds of oxygen. Your body combines it with the carbon in your cells, and you exhale 3 pounds of CO_2 and produce 1 pound of water.

By most ways of reckoning, our biological processes don't add to the net CO_2 in the atmosphere. The logic is that the crops we eat would have decayed and emitted the same CO_2 as they do when we eat them. In other words, the CO_2 just takes the scenic route through our digestive and respiratory tracts to get back into the atmosphere.

So, should the carbon dioxide emissions from fossil fuels be viewed as waste? Absolutely.

If combustion of fossil fuels did not emit any CO_2, no one would be staying up late at night saying, "If I could only figure out a way to use fossil fuels to increase atmospheric CO_2."

But beyond that, the CO_2 from energy production (or human respiration) is waste because it is *itself* a potential energy source. In a world with zero waste, we would use all the CO_2 we produce or exhale to power other things. After all, CO_2 is the food that powers most of the life on this planet, and the biomass of all plants is about a thousand times that of all animals. Plants take CO_2, combine it with sunlight and water, and produce carbohydrates (an energy source) and oxygen.

Why shouldn't humans be able to use that process? Shouldn't we be able to build an artificial leaf that, instead of using the photovoltaic effect to generate power like current solar panels, would use photosynthesis to make liquid fuel? Many people are trying to do just that. It's called artificial photosynthesis, and as Bill Gates, who backs one company trying to achieve it, says, "If it works, it would be magical."

Carbon Mitigation

If we don't use carbon dioxide in artificial photosynthesis, how can we reduce its prevalence in the atmosphere? One way is through direct air capture. This technology blows air across surfaces that are made for carbon dioxide to stick to. That CO_2 can then be stored away in the earth virtually forever—or turned into fuel and used as a power source to be reclaimed again later, in an endless loop.

Direct air capture is, as you might imagine, highly controversial. Alan Neuhauser wrote a piece for *US News* titled "Carbon Capture: Boon or Boondoggle?" in which a variety of scientists use terms including "sham" and "gimmick" to describe their contempt for this approach. They are quick to point out that you have to burn a great deal of fuel to run the carbon-capture plant, potentially releasing CO_2 back into the air.

Further, the promise of a machine that can pull carbon out of thin air might seem like a panacea to many, encouraging people to emit carbon with wild abandon. University College London professor Simon Lewis explains this concern. In a piece he wrote for *The Guardian*, he concludes, "While it is true that some negative emissions technologies are practically feasible at modest scales, this knowledge encourages both magical and mendacious thinking." Even if direct carbon capture could be done at scale, by one estimate it would require a quarter of all energy consumption in 2100 to power the machines.

However, there may be a path to using technology to remove carbon dioxide from the air for a cost of about $100 per ton. If this goal is achievable and scalable, the cost of removing *all* the new annual emissions of CO_2 from the air can easily be calculated. At $100 a ton, that's $100 billion per gigaton. Since we add about 20 gigatons of carbon dioxide to the air a year, it would cost $2 trillion annually. Against an $80 trillion world economy, that figure represents a 2.5 percent annual cost.

The challenge is that even if it's possible to remove carbon from the atmosphere (or prevent it from getting there to begin with) for $100 a ton, there is no mechanism to extract that amount of money from those who emit the carbon dioxide. As earth scientist and Stanford professor Ken Caldeira says, "The challenge is not in the magnitude of the task at hand, but that the interests and incentives to accomplishing the task are so widely diffused." Who pays the bill?

Removing carbon from the atmosphere doesn't absolutely require cutting-edge technology. There's one proven path to sequester carbon with which we're all familiar: the tree. Currently, the earth's three trillion trees store about 400 gigatons of carbon. If we add another trillion trees, we could store around 133 more gigatons of carbon, removing around 400 gigatons of CO_2 from the air.

A trillion trees seems like a lot, though, and critics argue that there simply isn't enough space for them in places that won't encroach on other needed human activities, such as farming. That said, Tom Crowther, a climate change ecologist at the Swiss university ETH Zurich, disagrees. His team used artificial intelligence to study satellite data and combined that data with thousands of soil samples his team gathered, to conclude that we actually have room for 1.2 trillion new trees. He states: "We are not targeting urban or agricultural areas, just degraded or abandoned lands, and it has the potential to tackle the two greatest challenges of our time—climate change and biodiversity loss."

Defenders of this approach point out that you can plant trees at scale for 30¢ each, meaning planting a trillion of them would cost just $300 billion, a low number compared to other types of solutions.

Critics argue that the situation is much more complicated, and of course there will be higher-order effects: Not only do we have to plant the trees, but we must maintain the forests we're creating, keep them from burning, and so on.

The debate hinges on just how meaningful such an approach would ultimately be. This vision of planting a trillion trees has been turned into an initiative overseen by the United Nations Environment Programme called, fittingly enough, "Trillion Trees." The project is already several billion trees along, working toward its lofty goal.

If you want to offset the CO_2 your activity produces, you can figure out how many trees you need to plant using some simple math. The activities of the average American release about 100 pounds of CO_2 a day. Over the course of an eighty-year life, an American will emit about 1,250 tons of CO_2. Given that a single tree will end up storing away half a ton of CO_2, planting twenty-five hundred trees should offset one American. Canadians and Australians are about the same, while most other prosperous countries' citizens emit a half to a quarter of that amount.

If you don't have a place to plant that many trees yourself, you can always buy carbon offsets—basically paying someone else to do an activity that prevents CO_2 from entering the atmosphere— for about $10 per ton of avoided CO_2. At that price, an American would need to pay $12,500 to offset their lifetime carbon footprint.

Ultimately, the main challenges to developing technologies that either mitigate carbon emissions or extract carbon from the air are financial, not technical. Carbon emissions are usually an economic externality. If I leave my car running on a hot day while I go shopping, so that it's still cool when I return to it, I've emitted carbon dioxide that causes some amount of economic harm, and I haven't had to pay for that economic harm. So if someone made a device that costs $1 that I could attach to my car to capture all its carbon emissions, I actually have no *economic* incentive to buy it. Even if it costs a penny, I have no economic reason to buy it. Likewise, in most parts of the world a factory emitting thousands of tons of CO_2 also has no economic incentive to mitigate its emissions, which is why in this arena governments rely much more heavily on regulation than markets.

Looked at another way, burning 1 gallon of gas produces 20 pounds of CO_2, meaning every 100 gallons creates 1 ton of CO_2. If the cost of removing that ton of CO_2 really can get to $100, then a $1 per gallon carbon tax where all the proceeds went to capturing CO_2 would effectively neutralize the effect of burning that gallon of gas.

Such an economic shift—charging $100 for every ton of CO_2 produced—would dramatically lower the amount of carbon emitted, as businesses would scramble to figure out ways to reduce their carbon emissions. Activities that emit a great deal of CO_2 would quickly become far more energy efficient as owners were forced to internalize the externalities of carbon emissions. Thus CO_2 emitters would make decisions based on the true societal cost of their actions.

The proceeds of the carbon tax could then be used to pull CO_2 from the air and sequester it, or as a rebate to the people hardest hit by the tax (which in this case would be the poor, who just saw the price of virtually everything they use go up). One recent survey suggested that support among the general population for a carbon tax skyrockets if the proceeds of the tax are rebated evenly to everyone in the population. Such taxes, known as Pigovian taxes, were originally popularized by British economist Arthur Pigou, and are a favorite among academic economists. Are they the solution to the CO_2 waste problem? We'll see.

It's Electric!

It's hard to imagine where humanity would be without language, arguably the most important technological advance in our history. The number two advance would be a bit more debatable, but the fact that we can generate and harness energy deserves serious consideration.

Worldwide energy consumption in all its primary forms (coal, solar, oil, etc.) totals 500 exajoules annually, 80 percent of which comes from fossil fuels. It's a colossal number, so big that any attempt to compare it to anything else gets bogged down in long strings of zeros. One exajoule is roughly equivalent to the energy contained in 170 million barrels of oil.

However, as much energy as 500 exajoules is, it pales in comparison to the 3.5 million exajoules that the sun rains down on our planet each year. This should give us some hope that however large our energy needs are, we have a whole lot more coming in from the sun to try to capture.

When we think about that number on a human scale, it's easier to put into perspective. Think of it this way: Your body runs on about 100 watts of power. On the other hand, the amount of energy consumed per person around the world is 2,500 watts. That means that what any given individual is able to do on their own is multiplied twenty-five times using harnessed energy. In some parts of the world, like in the United States and Europe, energy consumption is more like a hundred times that of the body. In other parts it's incredibly low, barely above the 100 watts humans ourselves are capable of.

Collectively, we pay around $8 trillion for all this energy. This amount represents about a tenth of world GDP and works out to about $3 a day per person. On average, humans can increase their power usage twenty-five-fold *for just $3 a day*. This is without a doubt the greatest bargain in the history of the human race.

We generate this energy in relatively few ways. About 80 percent of it still comes from just three fuels—coal, oil, and natural gas—in roughly equal proportions. The remaining 20 percent is obtained from other power sources including biofuels, nuclear, hydro, wind, and solar.

Some energy is consumed in the form of electricity, but not the largest portion; in fact, just 20 percent of all energy consumed is electric. That's less than what is used for transportation—everything from cars to trains to cargo ships to airplanes—which accounts for 25 percent. The balance is largely made up by industry as well as residential usage of oil and natural gas for heating and cooking. (Electricity is actually a pretty new technology. As recently as a century ago, fewer than half the homes in the United States had electricity.) Thirteen percent of the world's population today—over a billion people—don't have ready access to electricity. In many developing countries, including more than two dozen in Africa, a majority of people don't have easy access to electric power. In 2013, Ellen Johnson Sirleaf, the president of Liberia, pointed out that the amount of electricity used by the Dallas Cowboys' stadium during a home game was three times the amount that her entire country of over four million people got by on.

The world's total electrical production is about 25 trillion kWh. China uses the most, about 7 trillion kWh, and the United States uses 4 trillion. Divide that 4 trillion kWh by the population of the United States, and we find that per capita electric production is about 13,000 kWh. Of course, people don't individually use that much, because that number includes industrial and commercial usage. But on average, about 13,000 kWh are generated per person. (Interestingly, this number has been flat for over a decade even though we have more gadgets than ever, because those gadgets have become more energy efficient.)

How is that electricity generated? In the United States, according to the Energy Information Administration, about 4,000 kWh per person comes from each of coal and gas, about 2,500 kWh comes from nuclear, and another 2,500 is derived from renewables. Fossil fuels have always been the bread and butter of electrical generation. They have proven incredibly difficult to beat in terms of cost and energy density. The procedure for converting them into heat to boil water (the most common method of creating electricity) is simple and straightforward.

With regard to nuclear energy in the United States, all that power is generated by just about a hundred reactors operating at around sixty plants, the vast majority of which are over thirty years old. In fact, only one new nuclear plant has come online in the last decade, and few new ones are currently being built in this country. It's an open question as to whether nuclear generation will grow or shrink in the coming decades in the United States.

At its core—excuse the pun—nuclear energy still uses steam from boiling water to turn generators, but does so without fossil fuels. Nuclear energy is a polarizing topic. Most of us probably know most of the benefits and challenges of this form of electrical generation. But let's dive in a bit deeper to understand some of the fundamentals. A pound of coal, in theory, contains about 3 kWh of energy. However, our coal-powered electrical generation is only about 33 percent efficient, meaning we get only about 1 kWh of electricity from it. While this quantity may seem low, it's actually more efficient than the average car engine, which explains why it can be more efficient to centrally generate electricity to power a fleet of electric vehicles, even factoring in the energy loss that occurs as the electricity is transmitted and within the electric cars themselves.

If a pound of coal has 3 kWh of energy hidden away in it, what is the equivalent maximum energy in a pound of uranium-235, which is used to power fission plants? The answer: 10 million kWh. This is the allure of nuclear power.

That's the theory. In practice, nuclear plants "only" generate about 20,000 kWh per pound of fissile uranium. Nevertheless, that's still as much usable electricity as produced by 10 tons of coal. Of course, the 10 million kWh we could get in theory is much different from the 20,000 we get in practice, so where is the rest of the energy? Still locked up in the uranium. We can only pull less than 1 percent of the energy from the uranium out of it, and then the process to convert that energy to electricity is only about 33 percent efficient. Since the process uses such a small amount of the energy in the fuel, the weight of the waste uranium is virtually identical to the weight of the uranium that was used to create electricity. As a result, a large amount of nuclear waste is created when we use uranium-235 as an energy source.

Some reactors, on the other hand, can consume all of their nuclear fuel. The technology has existed since the 1940s and was developed as a hedge against a shortage of uranium. That shortage hasn't materialized since we've realized that uranium is as common as tin and not difficult to mine in significant quantities. Proponents of this kind of reactor technology (called a breeder reactor) suggest that with uranium extraction from seawater coupled with breeder reactors, it would be possible to power the earth for a billion years with inexhaustible clean energy. We could even use that energy to pull carbon out of the atmosphere, turning back the carbon emissions clock to whatever level we'd like. However, those suggestions are a topic for another book.

While the debate over renewable energy rages, one surprising fact about American electricity consumption is that in 1950 the United States generated a larger proportion of its power from renewables than it does today. While this bit of trivia might make for interesting cocktail party conversation, it's not particularly relevant to the current debate for a few reasons. First, the American population has more than doubled since 1950. Second, while the average American uses about 13,000 kWh a year now, in 1950 the number was about 2,000 kWh. Third, the average cost for 1 kWh of electricity is about a dime in current dollars; in 1950 it was about three times as expensive. The difference comes from hydroelectric power generation. Hydro, which generates about 6 percent of the United States' electricity today, generated 30 percent in 1950. And as reporter Tim Fernholz writes, "The era's lower demand for electricity meant that hydroelectric power coming from the mega-projects of the 1930s, such as the Hoover Dam or the Tennessee Valley Authority, could do a lot of heavy lifting."

Is it likely we will use technology to achieve a breakthrough in electrical generation? There are dozens, perhaps hundreds, of places for such an advance to happen. Perhaps we will figure out how to harness the huge amount of energy stored and released by the oceans every day as they rise and fall due to tidal action. There are winds in the upper atmosphere that blow at over 100 mph and never stop. Might we use those? The molten core of the earth hasn't cooled in the last four billion years and is still as hot as the surface of the sun. It's widely accepted that the amount of heat energy in the upper 6 miles of the earth's crust—a depth we can already drill to—contains thousands of times the amount of energy as is found in all of the world's known oil and gas reserves. So there is energy aplenty there, just waiting for a breakthrough or two. Others are looking at using the 400,000 tons of nuclear waste that we don't know what to do with as an energy source. And then there are techniques as exotic as molten salt solar power, where the energy of the sun is directed at a tank of salt, warming it up to its melting point (in excess of 1,000 degrees Fahrenheit). The dissipating heat is then used to power electrical generation after the sun has gone down, solving one of the more vexing challenges of solar power.

Physicist Freeman Dyson proposes a radical idea for energy: genetically engineering plants to make liquid fuel. There would be unlimited free energy anywhere the sun shines. This may not be as far off as it sounds. In late 2019, scientists at the University of Cambridge demonstrated an artificial leaf that can make syngas, a fuel used extensively today, out of just sunlight, carbon dioxide, and water. Currently most syngas is made from fossil fuels.

But perhaps the most elusive—and potentially least wasteful— of all technologies for electricity generation is nuclear fusion. Through fusion, humans would be, in effect, making and controlling small suns with very little in the way of unwanted byproducts.

One of the most promising efforts with respect to fusion is currently taking place through the creation of the International Thermonuclear Experimental Reactor in southern France. The reactor is expected to be operational in December 2025. If successful, these and other fusion experiments could provide a clear path toward a future like the one promised in science fiction novels, where clean energy produced with almost no wasteful byproducts is so inexpensive and abundant that it's not worth metering (which, the authors will hasten to add, was the original, failed promise of nuclear fission).

If such a time comes to pass, imagine what impact free energy would have just on the topics that have already been covered in this book.

Free electricity may sound like a pipe dream, but it wasn't too long ago that domestic long-distance phone calls—which are now too cheap to meter—were priced by the minute. In 2006, the Federal Communications Commission stopped tracking the cost, since most people weren't paying anything additional by that point. Bell's first telephone call took place in 1876, and long distance was free 130 years later. The first nuclear reaction to generate electricity took place in 1951; we would be willing to wager that it won't take until 2081 before we start thinking of electricity (or, more precisely, not thinking of it) as we do domestic long-distance calls.

Electricity Usage

How do we use all the electricity we generate? In the United States, electrical usage is split equally among homes, businesses, and industry. At home, half of the electricity we use is to heat and cool things—to power our air conditioners, heaters, water heaters, and refrigerators.

When we're trying to calculate waste, however, it's important that we know more precisely how energy is being used. Trying to figure out whether Netflix's shift in business models from mailing DVDs to streaming movies resulted in less waste is a vexing problem. In order to perform this specific calculation, we would have to know how much energy the internet itself uses. Those who have attempted to do so have achieved wildly varying results. A paper called "Electricity Intensity of Internet Data" by Joshua Aslan and colleagues tries to sort the whole problem out. They point out, "Existing estimates for the electricity intensity of Internet data transmission, for 2000 to 2015, vary up to 5 orders of magnitude, ranging from between 136 kilowatt-hours (kWh)/GB in 2000 . . . and 0.004 kWh/GB in 2008."

That's a huge range—136 is 34,000 times more than 0.004.

That would be like two people on *The Price Is Right* guessing how much an item is and one guessing $100 and the other guessing $3.4 million. So the difference is about as far from trivial as you can get. On the high end, at 10¢ per kWh, the price you're most likely paying for electricity, a gigabyte of information costs $13. These figures imply that downloading a 3-gigabyte movie uses $40 worth of electricity and generates 400 pounds of CO_2. On the other hand, if the lowest estimate is accurate, then downloading the movie uses not $40 worth of electricity but just one-tenth of 1¢ worth. The implications for waste are evident. (Note that the $40 or the one-tenth of 1¢ is not the cost on your electric bill, but the end-to-end cost for the energy used in the data transfer, including all the servers and co-location facilities between you and the data source.)

Why do the estimates vary so much? Two factors seriously complicate the situation. First, a huge proportion of the difference can be accounted for by the gains in efficiency of the equipment that powers the internet, which follows its own version of Moore's Law, with power consumption halving every couple of years. So the high-end energy costs always come from sources dating back several years, and the lower ones reflect more modern and energy-efficient equipment. This trend toward ever more efficient data transfer shows no sign of slowing down.

The second factor has to do with where the internet begins and ends. Do you count your computer if you leave it on all the time? Or just the time you are actively sending or receiving information? Or just the portion of your CPU that is handling data transmission, but not the parts that handle the keyboard and screen? Similar ambiguities exist all along the path from Netflix's servers to your tablet.

Even with these complications, it's highly likely that the cost of electricity needed to download a movie at present is much closer to the low end of Aslan's estimate. The movie you streamed probably used much less than a nickel's worth of electricity, with the largest variable being the device you watched it on. As a result, the shift from DVDs to streaming indeed represents a big reduction in waste.

Your iPad runs on electricity and it has to be charged to watch a movie. How much does it cost to do so? Answering that question is a good deal more straightforward. A smartphone battery holds about 2000 mAh at about 4 V, thus storing about 8 Wh of electricity. When we do the math, we find out that you can fully charge your smartphone twelve times for 1¢. Your iPad can be fully charged three times for 1¢, and your laptop, depending on the model, probably costs about 2¢ to fully charge. Watching the movie on any of these devices is still much less wasteful than the gasoline that would've been burned to deliver the DVD to your door—and you would still have to power your DVD player and television with electricity as well. (And on the topic of batteries, the manufacture of alkaline batteries—the kind you might have in your flashlight at home—takes more than a hundred times the amount of energy the batteries contain.)

The ubiquity of connected devices has resulted in the creation of waste as well. Take spam emails. An estimated 150 billion spam emails are sent each day. Even if we disregard the time it takes to deal with them, how much electricity is wasted on spam? The average size of a spam email usually contains about 25 KB or so of information. Knowing that, we can plug this data into the power usages just discussed. Using our low-end estimates, we can surmise that spam consumes 5 million kWh per year and results in the release of 3,000 tons of CO_2 into the atmosphere, roughly the same amount as ten round-trip flights from London to New York per year. Of course, the existence of spam causes waste in other ways as well: If there were no spam, your internet service provider wouldn't have to consume the electricity or the processor power needed to scan *every* email to see if it was a note

from a colleague or an offer for Fr33 V1agRA. And then there's
the time humans had to spend writing those detection algorithms
in the first place. Email spam has a large waste footprint.

What about other endeavors, like cryptocurrency? According
to Jonathan Koomey's paper "Estimating Bitcoin Electricity Use:
A Beginner's Guide," as of 2018, Bitcoin mining and usage consumes around 0.2 percent of all of the world's electricity. That's a
big number. If it's correct, it's roughly equal to the amount of
electricity used by all of the ten million inhabitants of Portugal. A
different study done by the International Energy Agency concluded that Bitcoin mining "is likely responsible for 10–20 Mt
CO_2 per year, or 0.03–0.06 percent of global energy-related CO_2
emissions." (The disparity between the percentage of electricity
used and the percentage of emissions is due to an estimated
75 percent of Bitcoin mining being done in places with cheap
renewable energy—looking at you, Iceland—so that it emits low
or no amounts of CO_2.)

Where should we start when trying to calculate where the true
waste is in the generation and transmission of electricity? Now
that we have more than a century of experience, it turns out these
calculations are pretty solid.

First, in the generation of electricity there's a good deal of inefficiency. As mentioned earlier, coal is about 33 percent efficient.
Natural gas is about the same, as are diesel generators. Solar is
less efficient, around 20 percent. Hydro, however, is highly efficient, in excess of 80 percent. Taken together, the energy efficiency of the entire grid averages 33 percent. In a world without
waste, we would get the same amount of electricity with just a
third of the fuel.

All of which leads to the conclusion that we're pretty far on the
waste side of the spectrum when it comes to electricity, especially
since another 4 percent of all electricity is lost in the form of heat
during long-distance transmission over high-voltage lines be-

tween the power plant and the city where the power is sent. Another 4 percent is lost in distribution from the low-power lines that run to your home.

But the waste doesn't stop there. Not even close. More energy is wasted as it's used in our houses. Computers are about 75 percent efficient; 25 percent of the energy they use gets converted to heat. Putting this waste in perspective, custom computer building company Puget Systems estimates that, per watt of energy consumption, *a gaming PC and an electric heater produce the same amount of heat.* Given that the waste heat from computers is rarely harnessed, that's pure waste. It also leads to interesting thought experiments about what we could do with all of the additional processing power we would have if we replaced our home heaters with supercomputers.

Beyond computers, though, appliances vary by efficiency quite a bit. In homes that still have them, the winner of the prize for least efficient device is likely the incandescent lightbulb, which converts only about 5 percent of the electricity it uses into visible light, with the rest going into heat. They generate so much heat—that is, waste—that when laws were passed restricting their use, some businesses tried to sell them as "space heaters" to skirt those laws. And truth be told, they're actually pretty good space heaters.

Devices that don't consume electrical energy vary widely in their efficiency. Home furnaces might be as low as 50 percent efficient or as high as more than 90 percent. Burning natural gas for heat is highly efficient—much heat is given off but very little energy escapes in other forms.

In the 1968 *Star Trek* episode "Return to Tomorrow," Captain Kirk tries to persuade his senior officers of the benefits of working with a certain group of aliens. He says to Mr. Scott that the aliens will be able to teach them how to power the *Enterprise* "with engines the size of walnuts." Scotty is incredulous but remarks that he didn't suppose "there'd be any harm in looking over diagrams on it."

Engines the size of walnuts? Would that even be possible? Easily. If 1 gram of matter—roughly the weight of a paper clip—were converted into pure energy, it would amount to 25 million kWh, or roughly the energy used by five Americans in their entire lives. An engine that could convert matter to energy like that on a regular basis could be quite small.

That we don't have such engines (yet) highlights just how far to the left we are on the waste continuum, and how much progress remains possible.

How did the disruption of the COVID pandemic affect electrical use? The answer is pretty intuitive: It lowered it. As businesses were shuttered and people telecommuted, electrical use fell to what is termed in the industry "Sunday levels." In other words, the pandemic was one long Sunday as far as power generation was concerned, with about 10 percent less electricity being used.

Sun to Table: Waste in Photosynthesis

Although the surface of the sun is a relatively balmy 10,000 degrees Fahrenheit, the core of it is immensely hot—27 million degrees of hot, to be precise. The sun is also enormous, more than a million times the size of the earth. Because it's so big, the gravity it exerts on itself is massive. And it's those two factors, its heat and the gravitational pressure from its mass, that together cause nuclear fusion.

The sun is mostly hydrogen, which you may remember is element number 1 on the periodic table, meaning its nucleus contains just one proton. However, if you cram enough hydrogen together under enough pressure and at a high enough temperature, eventually two hydrogen atoms will fuse, forming helium, which is element number 2. In the distant future, our sun will get hotter and hotter, fusing helium into heavier elements, then fusing those into still heavier ones, all the way up. Eventually it will explode, spreading heavy metals as it does. Our sun is a second-generation star—we know that because we can detect heavy metals in it, so it must be made of the debris of an even older star that ran through that whole cycle.

But right now, the sun is mostly hydrogen. When two hydrogen atoms fuse, energy is released in the form of both photons (aka light) and heat. The photons bounce around inside the sun for about four thousand years, according to NASA, until they break free and shoot out into space. Eight minutes and twenty seconds later, about one in every two billion of them hits our small speck in space, planet Earth. The rest disappear into the vastness of space, ending up as a twinkling light in the night sky of a distant planet. So from the perspective of humanity, 99.99999995 percent of the sun's energy is wasted.

Of the energy that does make it to earth, a third is reflected back into space by the clouds, the atmosphere, and the surface of the earth. But the amount that isn't reflected is still immense. About 10,000 exajoules of energy are absorbed by the earth's atmosphere, oceans, and land each and every day.

Of those 10,000 exajoules that come in every day, about 9,985 do very little beyond heating the surface of the planet and causing sunburns. Fifteen of those exajoules are crucial to the survival of the human race, since they're stored in plants, who take that energy and convert it into their own mass through a process called photosynthesis.

Let's pause for a thought experiment: Imagine you have a large pot of dirt and you plant a seed, which weighs essentially nothing, into that planter, and it eventually grows into a 100-pound tree. You then take the tree out and weigh the dirt left in the pot. How much is gone? In other words, how much of the tree's mass originated as the soil, and was transformed from dirt into plant mass? The answer is very little. Although it varies by plant species, on average only about 1 percent of a given plant's mass originally came from the soil. Let's call it 1 pound of our hundred-pound tree. So, where did the rest of the mass of the tree come from?

Half of the answer is probably pretty obvious: Like most living things, it's largely made of water. Most plants and animals are, which isn't surprising given how essential water is to life. If 50 pounds, on average, of the tree is water, and 1 pound is dirt, where did the other 49 pounds come from?

It came from the air. As every schoolchild knows, plants "breathe" in carbon dioxide and "exhale" oxygen. Carbon dioxide is made of two things: carbon and oxygen. One carbon atom, two oxygen atoms. Plants use the energy from the sun to take those carbon dioxide molecules apart, keeping the carbon and expelling some, but not all, of the oxygen—the O_2.

As the late-night infomercial host said, wait—there's more! Plants *also* use the energy from the sun to take liquid water (H_2O) and tear it apart, freeing some (but not all) of the oxygen. They then take some carbon atoms and some hydrogen atoms and combine them into a molecule of carbon, hydrogen, and oxygen. We call those molecules carbohydrates. Sugar is one of the most common of these carbohydrates, but there are others as well. And while some plants *can* use carbohydrates to make fats and proteins, carbohydrates remain mostly carbohydrates.

So, how much waste is there in this whole process? The plant takes in some energy from the sun and stores it in carbohydrates. But how much gets lost along the way?

It varies quite a bit. Many plants save just one-thousandth of the energy that falls on them. Crops planted by humans do much better, however. That makes sense, since we have bred them to absorb sunlight and make us rice and beans and so forth. Generally speaking, plants capture about 5 percent of the solar energy that falls on their leaves. (That's actually substantially worse than commercially available solar cells we have today, which achieve about 25 percent efficiency.)

Worse, of the energy crop plants absorb, only one-tenth of it is stored in the potato, bean, berry, or whatever part we eat. From our vantage point, the plant wastes all the rest on riotous living—

growing leaves and stalks, soaking up water, performing photo-synthesis, and so forth. Crop plants on average convert just ½₀₀th of the energy that lands on them into something for us. The rest is wasted. With regard to specific plants, science writer Gabriel Popkin calculated that one corn plant, which gives us 90 calories of corn, uses 27,000 calories of sunlight over its life. That means corn only gives us ⅓₀₀th of the energy it absorbs. Wheat is even worse, returning about ⅟₁₀₀₀th of the energy it takes in. Sugar-cane, on the other hand, is a high achiever, giving us back ⅟₅₀th of the energy that hits it.

When we stop to think about it, plants are remarkably waste-ful. We can build solar cells that use 25 percent of the sunlight that hits them. So why aren't plants anywhere near that mark?

To begin, half of the energy is lost right away, since plants can only use part of the solar spectrum. They can't absorb green light, so they reflect it—which is why they appear green to our eyes. (Solar panels, on the other hand, absorb all wavelengths, so they appear black.) Over half of what plants do absorb is lost in the inefficiencies of the photosynthesis process itself. Half of what remains gets used for respiration. Finally, the plant has what in a business context would be considered "high overhead." It has to grow a lot of stalk, leaves, and so forth, and can only store a tiny portion in the edible part that we consume.

Although plants are awfully inefficient, there may be a way to dramatically improve the yields of most of the foods we consume—more so than by just adding fertilizer. There are actu-ally three kinds of photosynthesis, of which two are important for our needs. First is C3, which is evolutionarily older and less efficient, and C4, which is a more recent development and is used by crops such as sugarcane and sorghum. If we can figure out a way to apply C4 photosynthesis to a crop like wheat, the areas where it could be grown would expand, as would yields. C3 plants have naturally evolved into C4 variants more than fifty times already, so it seems like a relatively common adaptation that with some human ingenuity might be helped along. (The third kind, CAM, is used by cacti, pineapples, and orchids, and only accounts for a very small percentage of all photosynthesis.)

Of course, it takes more than sunlight to grow crops for human consumption. It also takes labor to do all the growing and harvesting. In addition, there's machinery, water for irrigation, and fertilizers, as well as herbicides, pesticides, and fungicides. All of these inputs vary depending upon circumstances, but credible estimates put all of these factors together at a cost of about half the value of the crop. The math works out, in broad terms, to a cost of about a penny per 100 Calories.

As we progress toward a world that has no waste whatsoever, there are countless opportunities that depend on making better use of the energy that originates in our sun. The next few chapters will examine how humans currently squander so much of that star's bounty—and suggest what might have to happen in order for us to do better.

Food as Fuel

How does the food you eat become the energy you need to live your life? How much waste is there in your body's basic operation?

To tackle these questions, there are two terms we have to get straight: "calorie" and "watt."

What exactly is a calorie? We covered this earlier, but let's recap. The word is a bit tricky, since in common usage it can have two distinctly different meanings. You may remember from chemistry class that a calorie is the amount of energy it takes to heat 1 cubic centimeter of water (weighing 1 gram) 1 degree Celsius. So far, so good.

But if those calories were the same calories you see on the back of a food package, you could offset an entire sleeve of Fig Newtons by drinking a large glass of ice water and letting your metabolism raise that liquid to body temperature. Were that the case!

No, the chemistry class calorie is a "lowercase-c" calorie. The food label, on the other hand, shows "uppercase-C" calories, or kilocalories—a unit of measurement a thousand times greater than the lowercase-*c* calorie. A Calorie is the amount of energy it takes to raise a *kilogram* of water 1 degree Celsius.

If you drink a kilogram (about a quart) of water that's at a temperature of 2 degrees Celsius (35.6 degrees Fahrenheit), your body will expend 35 Calories warming it up to your body's 98.6 degree Fahrenheit temperature. A single Fig Newton has about 50 Calories, so eat them guilt free—if you're willing to drink a quart and a half of ice water after each one.

Calories are fixed units of energy. They are discrete measurements, like cups of flour or pounds of butter. They exist independent of time. Whether you increase the temperature of the kilogram of water by 1 degree over a minute or a millennium, you've still only put in one Calorie of energy. Whether you boil a pot of water with a match or with a blowtorch, it will take the same amount of *energy*. For practical purposes, it's often important to know how *fast* Calories are being transferred.

Enter the concept of "power." A blowtorch has more power than a match, since it can apply the energy needed to boil your water in much less time. Another way to think about it is to think about water flowing through a hose. Energy is the water; power is how much water comes out of the end in a given time.

Power is measured in units called watts. By reintroducing the time element, you can easily explain how much energy has been used in a given time period—say, an hour. A 1-watt lightbulb left on for an hour uses 1 watt-hour of energy—but so does a 60-watt bulb left on for one minute.

Getting back to powering our bodies: As luck would have it, the math is pretty simple because one Calorie and one watt-hour represent almost the same amount of energy. If your body plows through 100 Calories an hour, your body uses about 100 watts of energy—the same amount used by a bright incandescent lightbulb left on all the time. In a very real sense, if your body didn't waste any energy, it could take that 50-Calorie Fig Newton and turn it into enough energy to run that lightbulb for thirty minutes.

So how much energy does your body need just to keep you alive without doing much of anything else? If you basically slept all day, how much energy would you need? That level of energy is called your basal metabolic rate (BMR)—the energy needed to keep your heart pumping, your brain operating, and your body warm, for twenty-four hours. Your BMR declines as you age, which is one of the reasons it becomes harder to maintain a healthy weight as you grow older. For our rough purposes, let's use a nice round number, a BMR of 2,000 Calories, which is the value for a six-foot, 200-pound, thirty-year-old male.

Those 2,000 Calories *just* keep you alive, however. You have no energy to spare. So how many do you need to do something like mow the lawn with a power mower? If you have a largish yard, the size that takes about an hour to mow as you walk, you will burn 350 Calories, the amount you might get from an average hamburger. Those 350 Calories come from three different sources—carbohydrates, proteins, and fats—all of which are present in the aforementioned burger. Plants are (mostly) carbohydrates, while the edible parts of animals are largely made of protein and fat. Fat, with nine Calories of energy per gram, has a much greater energy density than either proteins or carbohydrates, which each weigh in at four Calories per gram.

Earlier we saw that plants take in water (hydrogen and oxygen) and carbon dioxide (carbon and oxygen) and create a carbohydrate, which contains some of each of those three elements. When they do, there is some spare oxygen left over and the plant gives it off as a waste product. This entire process is what's known as an endergonic reaction, which means that in the process of rearranging the atoms, some energy must be absorbed by the molecule from elsewhere. The plant gets the energy to do that from sunlight. So glucose is a hydrogen, oxygen, and carbon molecule with excess energy from the sun literally stored in the chemical bonds of the molecule.

How do humans make use of that energy? First, we eat some food, which is broken down by various digestive processes and is then sent to all the cells of your body. All energy in the human body is consumed at the individual cell level. You really do feed each of your cells individually. Within your cells are structures called mitochondria, which take pieces of glucose and break them back down into carbon dioxide and water, which is what the plant originally used to make it.

The plant stored energy when it made that molecule, so breaking the molecule apart releases that same energy, which becomes a new molecule called adenosine triphosphate, or ATP. ATP is what powers all living things, plants and animals alike. It is nature's ubiquitous power source.

Your cells then send the carbon dioxide they got from the glucose over to your lungs and the water to the kidneys, both to be expelled as waste.

The creation of glucose by plants is quite inefficient. And as you might expect, humans are imperfect machines as well. The conversion of glucose into water, carbon dioxide, and ATP also involves waste. About 40 percent of the energy in the glucose gets stored in ATP, while the other 60 percent is released as body heat.

Is body heat waste? Not exactly. Keeping your body warm inhibits the growth of thousands of harmful fungi that plague cold-blooded animals. Microbiology professor Arturo Casadevall estimates that the optimal temperature for maximizing the antifungal properties of heat while minimizing the costs of producing that heat is 98.06 degrees F, so our bodies aren't so inefficient in this regard.

Of course, we still radiate heat. Not all of the heat we produce stays within us; thus, a bit of it is wasted. Humans are pretty poorly insulated—our skin is only about 10 degrees cooler than our insides. But better insulation for humans would also come with some cost—and evidently, that cost hasn't been worth it for evolution to induce our species to pay the price.

The amount of heat each of us radiates could power a small laptop computer—if we could figure out a way to harness it. As it stands, we don't have any good way to get that energy back. Doing so isn't a technological impossibility; Seiko once released a watch that managed to run on the one-millionth of a watt radiated by the human body. And although there are a variety of efforts to make power-generating wearables, even the most ambitious are trying for fractions of a single watt.

What about feces and urine? Are they waste? Not in and of themselves, any more than a banana peel is. What matters is what we do with them. If they are recycled into energy-producing products such as methane, as much solid waste is, then it isn't really wasted. Even urine contains usable amounts of phosphorous, potassium, and sulfur, minerals we would otherwise have to mine. If you happen to live in a part of the world that lacks modern sanitation, however, then it is waste, and highly toxic waste at that. On the bright side, a United Nations think tank suggests that in aggregate, human bodily waste is worth almost $10 billion annually, and if the relevant technologies could be improved, it could be used to generate a significant amount of power in the poorest parts of the world.

Now we finally have a complete picture of where the energy comes from that enabled creation of the hamburger that allowed you to mow your lawn. The sunlight came to the earth and grew the plants, which were eaten by the cow, which we ate. We spent it pushing the mower, thinking, running our body, and radiating heat.

If we step even further back and look at the end-to-end story of where that energy first came from in the hamburger, it's mind-boggling.

To make the bun, the sun first had to emit energy equal to twenty thousand Hiroshima atomic bombs. That's a big number to us, but virtually nothing to the sun. Of that twenty-thousand-bomb energy, one part in two billion landed on earth, then one part in a thousand was converted by the plant into the wheat needed for the burger bun. In addition, to make that bun we needed 50 gallons of water and about 1¢ worth of labor, pesticides, and fertilizer.

To make the beef patty, the sun needed to emit 100,000 atomic bombs of energy. Of that energy, again, one part in two billion landed on earth. And of that, one part in a hundred became the grass our cow ate over the course of two years (grass uses energy more efficiently than wheat).

In addition, our burger required 500 gallons of water and about 25¢ worth of labor, materials, and incidental costs. Growing the meat resulted in 12 pounds of carbon-dioxide-equivalent greenhouse gases and 35 pounds of manure. To transport, process, refrigerate, and cook our burger took the energy from a pint of gasoline (or its equivalent in fuel) and produced 2 pounds of carbon dioxide. On top of it all is another 20¢ in miscellaneous costs, including human labor.

When you think about it, it's extraordinary how much has to happen to make one hamburger, and to give you the energy to mow the yard one time.

So, how much of the process is waste?

All of it. But why?

At the far end of the continuum, the one with no waste, we would simply absorb our energy at 100 percent efficiency from the sun. Or we could have mini fusion power plants inside of us, turning matter into energy directly in proportion to its mass and the speed of light squared. Alternatively, we could have solar collectors that stored it for us, beamed it from the collection point to wherever and whenever we needed, and allowed us to simply draw it from them.

The numerous changes in state as energy travels through space, then into plants, then into animals, then into us results in immense losses. Imagine taking $100 to the airport and changing it into euros, then the euros into yen, then yen into British pounds, and so forth. Pretty soon you'd end up with just a handful of change because of the commissions you have to pay on each step.

Could humans, in theory, cut out the intermediary (or some of the intermediaries) and just absorb our energy from the sun? Could we photosynthesize while still remaining recognizably human? Well, yes and no.

On the yes side, there is an aphid that stole a gene from a fungus and incorporated it in its genetic code in such a way as to enable it to photosynthesize. It can take sunlight and turn it into ATP, the same fuel that runs all life, including ours. And the yellow-spotted salamander, in its youth, has a symbiotic relationship with a type of algae that lives in its cells, performs photosynthesis, and shares the energy with the lizard. So it's not beyond the realm of possibility that with a little genetic nip and tuck we, too, could become at least partially solar powered, although we would likely need to turn green in the process.

However, in the no column, we just aren't physically cut out for photosynthesis. Humans are basically shaped like telephone poles. Our surface area is about 20 square feet. And while the average sunlight hitting the earth is about 15 watts per square foot, we can't expose all 20 square feet of us toward the sun at the same time. The best case might be half of that area. But at high noon with the sun overhead, it would be much less.

Let's assume that when the sun was out, we spent all our time lying down on a platform that always faced the sun, so half our body could collect sunlight. In that case, at perfect efficiency, we could probably get 150 watts, which is more than the 100 we need. The closer you live to the equator, the better, since that's where the sunlight hits the planet most directly. If you head north or south toward one of the poles, sunlight mostly hits at an angle, greatly limiting the amount of energy available.

What about if we tinkered with photosynthesis? What if we could get plants to absorb and use *all* the light that lands on them, then utilize all that to photosynthesize at 100 percent efficiency, and then use all their energy to make us food? If that were the case, we could sustain the entire world with a single farm the size of Maine, whereas we currently need land the size of the United States and Canada combined.

With regard to the inefficiencies of cattle and other food animals, if every calorie they consumed was perfectly stored as a calorie we could eat, then we would only need a pasture the size of Alaska, whereas currently we require pastureland the size of both the American continents to provide animal food for the world. But even in such a case, we would still have the inefficiencies involved in getting sunlight turned into grass for the cow to feed on.

As mentioned earlier, one solution is to directly manufacture the food we eat, especially beef, which is extravagantly inefficient. This idea is not new. In fact, Winston Churchill, in a 1932 essay called "Fifty Years Hence," wrote:

> We shall escape the absurdity of growing a whole chicken in order to eat the breast or wing, by growing these parts separately under a suitable medium. Synthetic food will, of course, also be used in the future. . . . The new foods will be practically indistinguishable from the natural products from the outset, and any changes will be so gradual as to escape observation. . . . If the gigantic sources of power become available,

food would be produced without recourse to sunlight. Vast cellars, in which artificial radiation is generated, may replace the cornfields and potato patches of the world.

Churchill got the idea pretty much dead on, although clearly his fifty-year estimate was way too optimistic. In theory, the amount of energy you need to create beef from scratch from cow stem cells would be much less than growing an entire cow for two years, and wouldn't require as much water nor produce methane, carbon dioxide, or manure. We are still in the early days of direct meat production now, but it will almost undoubtedly happen. The meat and poultry industry measures in the hundreds of billions of dollars in the United States alone, so there are plenty of economic incentives to drive the process forward.

Another option for less waste is to 3D-print food. In theory, the process for doing so is also straightforward: Just load up a 3D printer with tubes of protein, glucose, and fat, and press the button. At first glance, the process might not look like a net gain of efficiency. After all, you still have to use inefficient techniques to get the ingredients. But if we do it right, the things we regard as waste products, such as leaves and peels, can be turned into raw material to "feed" the machine. The protein tube may be full of powder from ground-up insects.

While the concept may not seem particularly appetizing, the proof will quite literally be in the pudding. Food is, after all, a trillion-dollar industry in the United States alone, so the incentives exist to transform it—especially as we move from a planet that currently must support eight billion humans to one that could, in theory, support many more, or support the same number with less of an impact.

PART 4

The Philosophy
of Waste

Wasting Money

In 2019, Oxfam International reported that the richest twenty-six people on the planet own more wealth collectively than the bottom half of the world's population. While the details of that assertion might be controversial, no one denies that there are dramatic disparities both in wealth and income throughout the world.

In 1912, Italian statistician Corrado Gini developed a method for measuring this inequality on a scale that ranges from 0 to 1. A society where all wealth (or income) is completely evenly distributed will have a Gini coefficient of 0; one where a single person has it all and everyone else has nothing has a Gini coefficient of 1. The World Bank has estimated that with respect to income, the world's Gini coefficient in 2013 was 0.625.

What does that mean? In general, it means that if you earn more than $40,000 a year, you are in the top 1 percent of global earners. Consider also the analysis of economist Branko Milanovic, who discovered that in 2011, the average income of the bottom 5 percent of households in the United States was higher than that of two-thirds of the planet, including the top 5 percent of households in India. In other words, the *poorest* twentieth of the United States earns more than the *richest* twentieth of India, even though India has the third-highest number of billionaires in the world, after the United States and China.

In the United States, it's almost a national pastime for people in different income groups to regard the spending habits of their counterparts with an air of disapproval. How people spend their money takes on an ethical dimension; to many, wasting money is not just imprudent, it is actually immoral. Many regard their own use of funds as soberly judicious while looking at others as akin to drunken sailors on shore leave.

In late 2019 it was reported that Baltimore Ravens quarterback Lamar Jackson had gifted some of his teammates Rolex watches for the holidays. Hearing this, a fan was disturbed by what she considered wasteful spending and took the time to write a letter to the *Baltimore Sun* suggesting that the money would have been better spent on charity. The hourly employees at the Rolex factory might not necessarily agree.

At least a luxury watch can tell you the time; one wonders how wasteful that fan from Baltimore would view other products designed for the rich that serve less obvious functions. For instance, if you happen to have $425, you can buy a pill full of gold leaf that you swallow in order to bedazzle your bowel movements—that is, to make your feces sparkle. Meanwhile, half a dozen countries get by with a per capita income less than that same $425 a year.

There's a pen on Amazon that sells for around $60,000. The three hundred or so reviews of the pen include such tongue-in-cheek comments as "When I run out of actual peasant's blood to write with, the red ink with this pen is an acceptable substitute" and "An unexpected value for the price" and "How can you afford NOT to buy it?" Famed jeweler Tiffany & Co. sells a sterling silver dog dish, which it says is crafted from "the highest-quality materials." And while the dish is indeed lovely and serves a function, it costs $1,800. Tiffany will also sell you a sterling silver ball of yarn—for $9,720. You would be hard pressed to find a better silver ball of yarn at any price.

Should silver not be to your liking, high-end retailer Neiman Marcus once offered for sale a 14-karat-gold-plated Slinky. The phenomenon is not limited to Americans, either; Dutch online retailer Ooms sells two dozen 14-karat-gold-plated staples for €59, suggesting that "you can also impress your boss by turning in your next report decked out in gold. He will feel you." Perhaps, but the company controller who approves your expense reports may not.

There's an episode of *The Simpsons* where Bart befriends a rich kid and goes over to his house, which is furnished with every cliché of the wealthy imaginable. Bart sees what he assumes is a poster of the boy's favorite American football player, Joe Montana, hanging on a wall. Upon closer inspection, it isn't a poster but a recessed area in the wall where Montana himself is paid to stand, holding a football in the throwing pose.

You may think that story is mere hyperbole. But in a life-imitates-art moment, a hedge fund manager once reportedly paid Kenny Rogers $4 million to perform "The Gambler" over and over as background music at his birthday party. (The hedge fund manager was subsequently convicted of conspiracy and securities fraud and sentenced to eleven years in prison and a $150 million fine. Evidently, he knew neither when to walk away nor when to run.)

It would be tempting to classify the Mercedes that may or may not have once been owned by Prince Alwaleed Bin Talal Bin Abdulaziz Al Saud as wasteful. In addition to its mink interior, the car is reportedly encrusted with three hundred thousand Swarovski crystals and is purportedly valued at around $5 million. Reports indicate that the car's owner charges $1,000 for merely *touching* the car.

Would it be a waste to spend $1,000 to touch a car? How about spending a few bucks on a painting that's simply a solid blue background with a vertical white stripe down the middle? What if, instead of a few bucks, it cost $44 million? In 2013 someone plunked down that cash for a 1953 canvas by abstract expressionist artist Barnett Newman.

What about the $2 million in cash that the Medellín cartel leader Pablo Escobar burned to provide heat for his daughter, who was suffering from hypothermia, according to the drug lord's son, Sebastián Marroquín? This last "expenditure" may seem more understandable; who wouldn't use whatever resources they had at their disposal to save their child from dying?

Are opulent purchases wasteful? It sure is tempting to say yes, but we can't. Trading $1 to listen to a song on a bar jukebox because you want to isn't actually that different from trading $425 to make your poop glitter because you want to. It's tempting to judge the latter, but presumably the person with $425 cannot think of anything on the planet they would rather spend their money on than a pill that gilds their feces. If that's the item that gives them the most joy, who are we to judge? We can suggest that an expenditure like this is unwise or imprudent (and more than a little bit gross), but we can't actually call it waste, as much as Puritan sensibilities might suggest we should.

It's like in the movie *Jackie Brown* where Ordell Robbie, played by Samuel L. Jackson, warns the character Melanie that smoking too much marijuana would rob her of her ambition. She replies, "Not if your ambition is to get high and watch TV." In our case, if your ambition is to have sparkly poo, a $425 pill might actually be the least wasteful way to get it.

It's all well and good to criticize the expenditures of the rich, who in some cases quite literally have money to burn. And it's not as if any wealth is actually being destroyed when a hedge fund manager hires a country singer to perform—that money merely gets transferred to the singer, who could choose to spend it on something he found valuable, from donations to the poor to plastic surgery.

And even when the money truly *is* destroyed, as in the case of Escobar, no real wealth is lost. When your rich Uncle Moneybags lights his cigar with a $100 bill, yes, he reduces his own net worth by $100. But in destroying that money, an economist would suggest (and George Mason professor of economics Alex Tabarrok *does* suggest, on Quora) that the only true waste is the cost of the paper itself: 14.2¢, according to the Federal Reserve. The $100 the currency represents causes deflation across the economy when it goes up in smoke, slightly lowering prices on all goods and services. In effect, destroying a $100 bill makes all the other currency in the economy together worth $100 more—resulting in net waste of near zero.

But enough of wealth-destroying plutocrats. What about waste among people who perhaps can't afford it the way millionaires and billionaires can?

Take lottery tickets, which are disproportionately purchased by people in the lowest fifth of the income distribution. Middle- and upper-income people often look scornfully on poorer people who spend their limited funds on these tickets, snidely referring to them as a tax on people who are bad at math.

But boy, are lotteries popular! Americans spend more money on lottery tickets than they do on music, going to the movies, books, video games, and tickets to sporting events . . . combined. Combined!

Since the odds of winning the lottery are effectively zero, have those folks wasted their money? No more than the person playing the song on the bar jukebox. Most people don't see any real tangible path from their current financial situation to wealth and riches. There aren't many Jed Clampetts shooting guns into the ground, striking oil, and winding up with prime-time sitcoms.

Nor do most people expect Ed McMahon to show up at their front door carrying an oversized check, especially since he moved on to that great talk show in the sky back in 2009. No, for most people, great riches are not likely to happen. However, for the princely sum of just a few dollars, anyone can buy a lottery ticket and spend a couple of days daydreaming about what they would do with all that money. Would they take the lump sum? Of course. What would they buy? Whom would they give extravagant gifts to? In front of which of their frenemies' houses would they drag-race their Ferrari? Perhaps they would follow the lead of Jonathan Vargas who used part of his $35 million in winnings to create *Wrestlicious TakeDown,* a TV show with female wrestlers. All of this imagining, with the added hope that maybe, just maybe it might come true, *has* to be worth more than $2 in entertainment value. At $10 for a movie ticket, the lottery ticket looks like a bargain.

But then you have to put a price on the disappointment of not winning. If the ticket cost $2, the value of the daydreaming was $5, and disappointment came with a $2 cost, then they're still ahead of the game.

And then there's the chance that they actually win! "Someone has to" seems to be a universal refrain. In Roald Dahl's *Charlie and the Chocolate Factory,* the title character is holding a chocolate bar, about to open it. The adults in the room are downplaying his chances of getting a golden ticket, to help prepare him for the disappointment that they "knew" was coming. Charlie has this thought: "But there was one other thing that grown-ups also knew, and it was this: that however small the chance might be of striking lucky, the chance was there."

Lottery winners supposedly fare poorly after winning, but this isn't exactly true. There was a 1978 study by H. Roy Kaplan that seemed to show that winning the lottery didn't make people happy, but subsequent research seems to paint a more complex picture. For instance, a 2004 paper titled "Work Centrality and Post-Award Work Behavior of Lottery Winners" by Richard D. Arvey, Itzhak Harpaz, and Hui Liao found that most lottery winners keep working, often at the same job, and almost never light cigars with $100 bills. Close neighbors of lottery winners, on the other hand, often try to keep up with their newly affluent neighbors and increase their spending and speculative financial investments.

Enough about the rich and poor—what about those in the middle? How about middle-income people and how they deploy money? What's wasteful? After all, there are 800 million hungry people in the world, and Americans and Europeans spend enough just *on cosmetics* to feed them all. Sixty percent of the world is still without indoor toilets, while Americans spend half a billion dollars a year on Halloween costumes . . . for their pets.

Obviously, there isn't a cause-and-effect relationship here. Someone in an impoverished country doesn't lack a toilet because an American bought a hot dog costume for a dachshund. But someone looking at the world as a whole might ask whether all that makeup and those pet costumes aren't wasteful in a world struggling with scarcity.

Again, we can't call it waste. The person buying a miniature whiskey barrel to put under the chin of their St. Bernard is increasing their utility—that is, their happiness—by buying it. They would rather have that whiskey barrel than $40, so it was a good trade for them. They're happier, and the people who made and sold them the item are happier. Perhaps we wish we lived in a world where donating that $40 to feed starving kids gives people more happiness than a pet costume. But evidently, we don't—to the tune of half a billion dollars.

It's hard to judge anyone for the choices they make with their money. One can't tell people they should prefer buying museum tickets instead of lottery tickets, any more than one can tell them they should prefer the color green over the color blue. So, how do we identify situations that waste money?

There are four ways money can truly be wasted. The first occurs when you don't get the value you expect in a transaction, due to either fraud or misinformation—that is, when your expectation of the transaction is not met. Years ago, national ads appeared in magazines selling a solar-powered clothes dryer for $49.95. Recipients who purchased the dryer received . . . a length of clothesline. If you've ever ordered a hamburger at a fast-food place based on the photograph hanging above the menu and then been disappointed when your actual burger looks nothing like it, you know this feeling. And if there's a way to train Sea Monkeys to do tricks, the authors haven't figured it out. That money is wasted.

The second occurs when there's a misunderstanding about the nature of a transaction you enter into. When some "fine print" provision of an agreement reduces the value you think you're getting in a deal, it's waste. In contract law there's a notion called a meeting of minds. It states that you can't have a binding contract when parties don't have the same understanding of the nature of the agreement. If you didn't understand that by buying a new car you were also agreeing to pay the dealership for undercoating, that $500 is wasted. If you didn't realize that the only time you could use that timeshare property is when your kids are in school, well, it's a waste.

The third situation occurs when you enter into an adverse transaction unwillingly. When some mysterious feature gets added to your cellphone plan that you didn't even know was there, that's wasted money.

Finally, money is wasted if you're deemed mentally unable to enter into a transaction. This is more common than it sounds and is the reason that children often can't sign contracts or engage in certain activities. If someone is drugged and gets an image of

Betty White tattooed on their biceps in Acapulco, that's wasted money.

It's actually possible, by the way, to waste money in transactions that *do not involve anyone else*. In fact, this is how most money is wasted. Consider the case of James Howells of England. He worked in IT and managed to acquire 7,500 Bitcoins, which were worth very little at the time. The laptop they were stored on became outdated, so he broke it up for parts, putting the hard drive in a drawer in the highly unlikely case those Bitcoins ever became worth anything. Along the way, the hard drive was inadvertently sent to the landfill. Oh, and the Bitcoins became worth over $100 million. That's wasted money any way you slice it. Howells is trying to get the town to dig up the landfill—wouldn't you?—but it's not likely to happen. Unlike a certainly destroyed $100 bill, however, the uncertainty of whether those Bitcoins might ever show up means the other Bitcoins in circulation don't increase in value by the same amount.

To be clear, there actually *was* a transaction here. At some point Howells decided, "I will trade this old hard drive away to get a clean drawer." But he found himself in the first type of waste situation mentioned earlier: not getting the value he expected from that transaction because of bad information. In this case, he didn't have the information that the drive at that time was already worth $100 million. Additionally, he also found himself in the second type of waste situation: misunderstanding the nature of the transaction itself. He thought he was trading away a worthless drive for a clean drawer, and such was not the case.

Similarly, in 1977 after a fire destroyed part of a Dublin observatory, the Apollo 11 moon rocks that had been given to Ireland were thrown into a landfill along with other burned material. The rocks were worth several million dollars, and their loss can only be called a waste, for the same reasons.

But we mustn't make the "error" category too expansive. What we're talking about is when money, or something that can be reasonably converted to money, is lost or destroyed through a fundamental misunderstanding of the transaction. That's a different situation than when events don't unfold as you expect them. If you buy car insurance and never have a wreck, your premiums aren't wasted money in this analysis. Indeed, you could have had a wreck, and you got full value from that insurance in the comfort of knowing that if you had a wreck it would be covered.

What about cases like medical error, the third-leading cause of death in the United States? Your expectation of the transaction wasn't that the doctor would kill you. If you die, your premature death would be considered a waste. However, the money you spent on the procedure wasn't wasted because you went into the procedure knowing there was risk involved. In this way, it's no different from the car insurance. (But if a doctor guaranteed that a procedure was 100 percent safe, then killed you during the procedure, that's waste, and it falls into more than one of our categories.) Similarly, when you bet on black on a roulette wheel and the ball lands on red, your bet wasn't wasted.

What of slips on icy sidewalks and auto accidents? The money spent to fix problems caused by such events isn't wasted, because we walk on icy sidewalks and drive knowing that these risks exist.

The upshot of it all is that when a mentally competent person who understands the nature of a transaction and possesses accurate information voluntarily undertakes that transaction, it's hard to say that money was wasted.

Wasting Time

We are obsessed with time. The folks over at the *Oxford English Dictionary* say that the word "time" is the most commonly used noun in the English language, with the word "year" third, while "day" and "week" both make a showing in the top twenty.

Our fascination with time reflects how much we value it, and by extension how we want to avoid wasting it. And we've worried about this for quite a while. Four centuries ago, in William Shakespeare's *Richard II*, the eponymous monarch utters the line "I wasted time and now doth time waste me."

But even Shakespeare was late arriving on the wasted-time bandwagon. Two millennia ago Seneca penned, "It is not that we have a short time to live, but that we waste a lot of it. Life is long enough, and a sufficiently generous amount has been given to us for the highest achievements if it were all well invested. But when it is wasted in heedless luxury and spent on no good activity, we are forced at last by death's final constraint to realize that it has passed away before we knew it was passing." They didn't call him Buzzkill Seneca for nothing.

One of the great equalizing facts of life is that everyone on the planet has exactly twenty-four hours in a day. Although we all individually choose to spend that time differently, let's look at how we collectively use those hours and try to sort through how much of that time may be wasted.

A logical place to start is with sleep, the activity that takes more of our time than any other. The average person sleeps, or tries to sleep, about nine hours a day. Over an eighty-year life, sleeping nine hours out of every twenty-four means we're asleep for a total of thirty years. Is sleep wasted time? Only to the extent it is avoidable and is not considered pleasurable. If sleep is necessary for survival or if a person finds it pleasurable, it's not waste.

While the pleasure value of sleep varies, scientists still don't understand physiologically *why* sleep is required for human life—but it evidently is. You can go longer without food than without sleep. The effects of modest sleep deprivation, such as staying up for twenty-four hours, mar your cognitive ability as much as being intoxicated.

Many sleep scientists believe you cannot train your body to need less sleep, and that doing so merely builds up a sleep deficit that manifests in poorer performance, which must be paid back at some point. Some argue that there's a more efficient way to sleep than in one block, and that our recent ancestors, just a few centuries ago, spent fewer net hours sleeping by alternating shorter periods of sleep and wakefulness—a practice known as bimodal sleeping.

If humans can go without sleeping, the world's militaries would like to know how. An article in the *New York Times,* for instance, points out that bristle-thighed curlews routinely fly as much as 6,000 miles without a stop as they travel from Alaska to the Marshall Islands. Assuming an average speed of 20 mph, that's more than thirteen days without sleep. DARPA, the research and development arm of the U.S. Defense Department, is studying birds to see if soldiers can consistently do the same. Not the flying part, obviously, but the staying awake part. Figuring out a way to juice soldiers is hardly new; the Third Reich's devastating blitzkrieg invasion of France was powered by amphetamines that kept their soldiers awake for days and extended how long they could march at a time. But amping up armies with stimulants goes back way before even the last century. Cocaine was widely used in armies during World War I, and other drugs have been given to soldiers at least as far back as ancient Greece.

There exist tantalizing accounts of humans who allegedly needed no sleep at all. Al Herpin, who entered into eternal sleep in 1947 at age ninety-four, claims to have spent the later decades of his life getting absolutely no sleep at all. Or consider Paul Kern, a Hungarian soldier who took a bullet to the head in World War I and is said to have never been able to sleep again in the subsequent forty years despite the use of hypnosis, sleeping pills, and alcohol. Other stories along these lines seem to have been widely believed and reported in their time but are seriously doubted by scientists today.

In modern times, the longest verified amount of time a human has gone without sleep took place in 1964 at Stanford University, under scientifically rigorous conditions, when teenager Randy Gardner spent eleven consecutive days awake. During that time, in which he was being closely monitored for signs of microsleep that might have escaped his notice, he maintained reasonably normal function, albeit with some clear evidence of cognitive decline. Afterward he slept for fourteen hours, then ten the following night, and was then back on track, evidently none the worse for wear. Well, maybe that's not quite the way to say it. He went on to suffer from debilitating insomnia, which he describes as "karmic payback" for his teenage stunt.

Certain diseases can interfere with sleep—in some cases, fatally. In an illness out of a horror novel, those who contract fatal insomnia see their brains invaded by prions, misfolded proteins believed to be able to transmit maladies. For the first four months of the disease, the unlucky patients suffer insomnia that leads to panic attacks and paranoia. For the next five months, they suffer from hallucinations. Eventually a complete inability to sleep leads to rapid weight loss, dementia, and eventually death. The whole process usually takes about eighteen months.

Although hardly as severe as in the case of fatal insomnia, accumulated sleep deficits take their toll on us all. Fatigue due to diminished sleep is considered a contributor to the cause of any number of disasters, from the *Exxon Valdez* to Chernobyl.

How, then, do we spend the other fifteen hours left in our day? Of the approximately seven hundred thousand hours each of us is allotted in our lifetime, we will each spend an average of ninety thousand of them working in an occupation. That may sound like a lot (and depending on the job you have, it might feel like even more). But, doing the math, you'll find that it only works out to a lifetime average of three hours a day. Remember, of course, to subtract out your youth and retirement years, weekends, vacations, and for some, those days when your boss isn't watching closely.

On average, people work eight hours a day, 225 days a year, for fifty years. By a conventional Western definition, it would seem that these hours are more or less the opposite of waste. If someone is paying you to do something, regardless of how mundane, irrelevant, or boring it might be at the moment, it must be worth something to them, and it creates work for you—so we can't classify any job where someone is paying you with their own money as waste.

What are we doing when we aren't sleeping or working? In the United States, the second-largest use of our time is actually . . . television. According to Nielsen, as recently as 2018 we spent four hours a day watching it. That's broadcast television in real

time, the same way the Pilgrims used to watch it. We're not talking time-shifting DVR or YouTube, just plain TV. And given that
nearly a quarter of that time is commercials, that's an hour a day
being told you have bad breath or are balding or that your car
isn't quite adequate. Multiply the numbers out over a lifetime,
and you're likely to spend well over two years of your life just
watching commercials. That doesn't count all the other ads you
come across, all the radio and billboards and internet ads. Is that
two years of life wasted time? Maybe not. If you value the shows
you are watching, well, they have to be paid for, and that's what
you are doing by watching those ads.

TV isn't even a majority of the media we consume. According
to the same Nielsen study we spend eleven hours a day consuming media, which includes reading, listening, and watching.
There's overlap here with the work time we discussed earlier. You
may be watching a video at work, which would result in the double counting.

Of those eleven hours, we spend three on our smartphones.
This particular part is a new phenomenon, and it accounts for
our rising consumption of media. Our behavior here has been
heavily modified over the last decade. Now many of us can't step
into an elevator without whipping out a phone to check email.
How else are we going to spend the eternity it takes to get to the
seventh floor? But again, we can't categorically count any of this
as wasted time.

What about time spent in the car? On average, drivers in the
United States spend about an hour a day and drive an average of
30 miles. Over the course of your life, the amount of time you
will spend waiting for red lights to turn green is measured in
months. Is all of that wasted time? Conceptually, yes. After all,
you'd ideally want to be able to step into a *Star Trek*–like transporter and just appear where you want to go, right? But trans-
porters don't exist, and we are willing to spend the time to get

wherever we are going. So we can't practically regard this time spent as waste until there is a better alternative.

What about time stuck in traffic? According to a study of commuters by the Texas A&M Transportation Institute, in 2014 "congestion caused urban Americans to travel an extra 6.9 billion hours and purchase an extra 3.1 billion gallons of fuel for a congestion cost of $160 billion." The same report notes that in order to be on time for an important freeway trip, on average drivers left half an hour earlier than they would have otherwise, to account for traffic. Is this time wasted? In the strict sense of the word, yes. But it's waste we choose. We could build more roads, mandate mass transit, cap the size of cities, and require carpooling, but we collectively choose not to. Evidently, we would rather wait in traffic than do these things.

Then there's . . . everything else. The average person spends an hour a day eating and another hour a day doing chores. Over a lifetime, each of those amounts to a couple of years of your life. Add in another year of your life sitting on the toilet.

We're dealing in pretty round numbers here, but if you're keeping score you will have noticed that all this activity, from sleep to work to media and all the rest, adds up to more than 24 hours a day. How can this be? It happens because we multitask, so things get double-counted. You may watch reality TV while in the bathroom or listen to music while doing the ironing. If you're a real overachiever, you can take a nap at work while a video is playing on your computer and score a three-fer.

So far, we haven't identified any ways to eliminate wasted time. Yes, you could theoretically take a helicopter someplace to avoid traffic. But for most people that solution is impractical. If you're making a voluntary choice, based on what's important to you, on how to use your time, it's hard for someone else to conclude it is a waste.

By extension, a person who spends all their free time playing video games and drinking beer isn't wasting their life. Presumably they're doing exactly what they want with their time. And further, going out and getting a better job would actually be a waste of time from that person's perspective, since they would be spending that time doing something they didn't really want to do. They'd rather be at home slacking.

An old joke about a fisherman and a banker has the banker on vacation in a small coastal village criticizing the fisherman's lack of initiative, as the fisherman works only a few hours a day to make just enough to feed his family. The banker explains that by borrowing money and working longer hours for twenty or thirty years, the fisherman would be able to retire. "Retire?" asks the fisherman. "What would I do if I retired?" And the banker responds, "Well, I plan to move to a small coastal village and fish for a few hours a day."

As a common adage goes, "the time you enjoy wasting is not wasted time." Henry Ford, however, would not have concurred. He offered, "It has been my observation that most people get ahead during the time that others waste." Michelangelo might have agreed with Ford. It's said that after his death a note was found written in the aged man's hand for his apprentice that simply said, "Draw Antonio, draw! Draw and do not waste time."

If any of this makes you feel bad, consider that the animal kingdom is full of time-wasting slackers, too. By one estimate, in a given ant colony, 3 percent of the ants are workaholics and never stop, about a third seem to do absolutely no work at all, and the rest work some and slack some. These percentages may not be all that different from a human colony. Ants also have behaviors that sure look wasteful. They stack their dead in intricate ways, or bury them in specific spots, only to later rearrange them or dig them up and move them.

If slackers who enjoy doing nothing aren't wasting time, can time actually be wasted? Can we shift down the continuum toward zero waste with respect to time? Absolutely. Let's consider a few specifics.

There's time you spend looking for something you lost. This may not sound like much, but consider how often you misplace your keys, your smartphone, the remote control, and your umbrella. But don't stop there. What about missing computer files, or that website you visited last month that you need to find again? Or that video you saw online? You know, the one with the guy wearing the blue shirt? How do you find that? How about a lost passport? W-2 form? Birth certificate? What about the combination to a lock? Lost passwords? Finding all of these just puts you back to where you were. And as Ben Franklin quipped, that lost time "is never found again."

Just how much time is wasted looking for stuff? This phenomenon hasn't been well studied, but if you misplace something every day and take five minutes to find it, then 100 days of your life would be spent looking for stuff you've lost.

In a similar vein, that sinking feeling you get when you forget to save a file and have to redo a bunch of work clearly could have been avoided. We waste time when we get lost, to be sure. And waiting in lines, such as at the DMV, feels like wasted time as well.

Being sick could count as wasted time if what you got was avoidable. Eventually, it's highly likely that technology will allow all sicknesses to be prevented or cured. But in the meantime, the colds and other illnesses you get feel like a waste.

What about ways to prolong your life, to add additional time? One candidate could be to take up a physical activity like jogging. Some older research suggested that you could increase your life span by three years by regularly jogging. But if you do the math, you would be spending those three years, well, jogging. Is that worth it? You decide.

Newer research suggests you can get a much better return on

your jogging misery. An hour of jogging a week, spread out over four days, works out to about six months of jogging over an adult life, and could add perhaps six years to your life span. Biking seems to yield a similar return. A high-stress life, on the other hand, takes away about that much time. Being happy is perhaps the best medicine of all.

But the easiest way to eliminate wasting time is through technology. We don't walk up the hill to haul water from the well anymore; we turn on the tap. Imagine household chores before electricity: washing clothes before washing machines, heating a home with wood, et cetera. While labor-saving devices at work haven't lessened the number of hours people on average put in at the office, time recouped with labor saving devices at home has been pocketed, so to speak, and we now have it for leisure. That's why we can consume media eleven hours a day.

Now, if someone can simply invent the transporter . . .

Wasted Human Potential

Since the beginning of the human race, we've spent most of our waking hours simply trying to survive. The quest for food has taken most of our time, whether in hunting and gathering or in farming. We've devoted time and energy to other essentials, such as shelter and clothing. The few remaining hours we've often spent on other survival-related activities such as self-defense. We went from making flint arrowheads to crafting bronze swords to manufacturing ballistic missiles, while our enemies did the same.

Survival, in short, was a full-time job. Scarcity was the governing principle of the world. There simply wasn't enough of the good stuff—food, medicine, money, education, leisure—for everyone to have it. This simple fact is the core foundational principle of all economic theory, from capitalism to communism and all points in between.

Over time, as we explored in the chapters about food and clothing waste, we gradually learned how to become dramatically more efficient and created technology that allowed us to produce more than enough to provide for our basic needs. We did this by expanding our abilities and increasing our productivity. We built a world that no longer required the collective effort of virtually everyone simply to survive.

What did we do with this surplus? A lot. We invented the arts, where some people could focus not on aiding in survival but in inspiring, delighting, terrifying, or eliciting some other emotion from their fellow humans. We formalized the concept of leisure, time set aside for activities that were not strictly necessary to live but from which we derived personal enjoyment or edification. And we created wealth, which meant that resources could be stored up for future needs or extravagant delights. Humans aren't unique in the way we deal with abundance; when nature delivers a surplus of nectar-producing flowers, bees just work that much harder and make that much more honey, vastly more than they could ever put to their own use. Of course, that's pretty much all they know how to do, so this workaholic lifestyle is understandable. But it's less so in humans—especially in those who, upon earning their first billion, rise early the next day to start on their second one.

With our time and our wealth and our leisure, we began to build a world with dramatically increased choice and self-determination. In prior times, if you were born a farmer—as virtually everyone was—you would grow up to become a farmer. You would grow whatever crops your neighbor grew, the same exact way they did.

Wherever you entered the world, you were almost certain to leave the world within walking distance. The religion you practiced, the foods you ate, the clothing you wore—all would be determined by your parents. Your spouse would probably be selected for you, and unless he or she had a physical impairment, you would have children. Your friends would virtually all be chosen from a small pool of people who shared your age, gender, race, social class, or geographic proximity.

Just think about that. In that world, what real choices would you be able to make? Except within prison-like narrow boundaries, you couldn't decide what time to get up in the morning or who would rule over you, or even how you would spend more than a few stray moments of your life.

It's interesting and simultaneously heartbreaking to think of all the Da Vinci–esque geniuses who have been born across the countless eons of human existence whose short lives were spent in a world with no opportunity and no choice. All the proto-Michelangelos whose gifts went no further than perhaps inspiring a quick smile from a family member over a drawing scratched in the dust. How many lived and died in a world that, day after day, month after month, year after year, offered only relentless toil just to survive?

Interestingly, by this book's definition of waste, we cannot say those lives were wasted. Waste, by its nature, is unnecessary, and in times past it was quite necessary that everyone spend most of their time surviving. We were too busy staying alive to worry about self-determination. That being said, it's still unfortunate that so many generations had to live lives so rigorous and regimented that they had little opportunity to achieve their potential. The would-be Marie Curies, whose gifts were suppressed due to their gender; the would-be George Washington Carvers, whose abilities never reached fruition due to the level of melanin in their skin; and the countless others who had virtually no agency over their lives and destinies due to arbitrary factors of their birth, point to a nearly unfathomable waste.

Things have changed since then. We live in a world where, in theory, more people should be able to develop their interests and abilities at least part of the time. "In theory" is the critical part of that last sentence. In practice, very few are still afforded this sort of self-determination.

Why so few? What limits our ability to achieve our maximum

potential? First and most significant is poverty. Half of the world lives on $4 a day or less. A quarter gets by on half that much. The world is still so poor that if you make $40,000 a year, you are in the top 1 percent of earners worldwide. And while the world has a great deal more income mobility than it used to, great wealth is still economically quite persistent across generations. According to *The Guardian*, in England 160,000 people out of a nation of 55 million—well under 0.5 percent of the population—own two-thirds of the land. And these people are largely descended from those granted tracts of land after the Norman Conquest a thousand years ago.

For the half of the planet making $4 a day, life is still much like that of their forebears who had little time for leisure activities. If you're an Amelia Earhart in such a place, you won't be taking flying lessons. If you're a Pablo Picasso, there isn't money for paint.

From poverty comes all kinds of other factors that hold people back. Globally, the poor don't get preventive healthcare like vaccines. A billion people don't have clean water, and three billion lack basic sanitation. From poverty stems malnutrition, the effects of which ripple through the world. From the half million kids who go blind each year for lack of a few cents' worth of vitamin A to the hundred million kids whose growth is stunted due to hunger, malnutrition is a blight on this planet whose time should have long ago passed. Hunger could be completely eliminated in the world for the annual cost of about $100 billion, a sum just under what we collectively spend on pet food.

Lack of money generally also means lack of education. Not just higher education but even the most basic education is disrupted in places that experience poverty. About one in seven adults on the planet is illiterate, using the definition of being unable to read at age fifteen or over. Lack of money means lack of medical care, too, meaning there are those with diseases and conditions, perhaps quite treatable, that rob them of their potential.

Then there are institutional factors that limit human potential. From sexism to racism to classism to nationalism, many paths are closed off to great swaths of the population based entirely on the circumstances they were born into. Next are the walking wounded. They are those to which life has dealt such terrible blows that they're forever impaired. They all need our help, perhaps for their entire lives. There are those who were abused, those who succumbed to substance abuse, those who suffered debilitating mental and physical injuries, those whose development was stunted by exposure to toxic chemicals and the like. Next to them are those who are robbed of their childhood by being conscripted into fighting in wars, not to mention those who lost their lives. Those wrongly incarcerated. Those detained or killed for simply speaking out or supporting the wrong political faction. And there are those whose lives are cut short or impaired by easily preventable accidents, criminal acts, or any of a thousand other causes.

Imagine a world where everyone had the opportunity to achieve their maximum potential. In the grand scheme of things, only a few elites have anything close to such power over their lives today. The exhortation that "you can do anything you set your mind to" is true only for a very few people. If you're one of the fortunate few, you can choose an educational path that probably involves a university, then select a profession. If a job isn't to your liking, you can get a different one—perhaps taking a lengthy period of time with no profession at all. If a dozen years into your career you decide you want to be a research scientist or a candlemaker, you can ditch the life you have and start afresh. Doing so may not be easy, it may involve debt or an investment of time, but it's within your reach. You can speak your mind freely without fear of persecution, pursue things that interest you, and expend energy over causes that excite you.

Now imagine a world in which *everyone* had that kind of

power over their own lives. Think of all the diseases we could cure if a hundred times as many people worked on each one. Imagine the art that would fill our lives and the great literature that would be produced. The gap between the world we have and this world we can imagine is the scale of the wasted human potential that occurs today. And that waste is so large it's truly beyond our imagination.

The gap between what could be versus what is may sound quite disheartening, but there's reason to have a great deal of hope. In virtually all places, by virtually every measure, things are getting better in terms of giving people more self-determination and agency over their own lives.

Every day there is more literacy, more self-government, more access to medical care, more individual liberty, more equality before the law, and rising standards of living everywhere you look. What originally took millennia to begin in the first place now can happen quickly. South Korea went from 22 percent literacy in 1945 to 88 percent in 1970 to 99 percent today. But as fast as these sources of waste are being eliminated, the process is still heartbreakingly slow, and puts the waste that occurs from ordering the wrong size sneakers from an online shop into dramatic relief. Next to the waste of lives, everything else pales.

Is Humanity the Fundamental Problem?

In the first *Kingsman* movie, the villain Valentine explains the rationale behind his plan to kill off a sizable part of the world: "When you get a virus, you get a fever. . . . Planet Earth works the same way: Global warming is the fever, mankind is the virus. We're making our planet sick. A cull is our only hope."

This same concept appears in *Godzilla: King of the Monsters* as well. Dr. Emma Russell explains to her husband and daughter why she is waking up all the monsters to wreak havoc on the human race. She recounts humanity's various crimes against the planet and then states: "Our world is changing. The mass extinction we feared has already begun, and we are the cause. We are the infection. But like all living organisms, the earth unleashed a fever to fight this infection: its original and rightful rulers, the Titans."

Likewise, in *Avengers: Infinity War*, Thanos gives his explanation for wanting to kill half of all beings in the universe: "The universe is finite. Its resources, finite. If life is left unchecked, life will cease to exist. It needs correction."

In *The Matrix*, the nonhuman Agent Smith explains what he

thinks is wrong with humans: "You move to another area, and you multiply, and you multiply, until every natural resource is consumed. The only way you can survive is to spread to another area. There is another organism on this planet that follows the same pattern . . . a virus. Human beings are a disease."

This trope appears over and over in popular culture. Granted, it's usually the villain who holds such a negative view of humanity. But because they have seemingly good intentions, such villains are sympathetically portrayed.

Here's the thing: If you wake up every morning and say "We are a virus" a dozen times, then repeat it a dozen more times before bed each night, eventually you will start to believe it, whether it's true or not. That's the cumulative effect of all of the stories that rely on this notion.

Humans generate virtually all the waste on the planet. So if you really want to help the planet, one could argue, why not build painless vaporization chambers that people can step into and—zap—cease to be? Would well-intentioned waste-haters line up around the block waiting for their chance to step into the death chamber, eager to do their bit for efficiency and/or Mother Earth?

You may scoff, but there's something called the Voluntary Human Extinction Movement, which makes the argument that the world would be better off without us. They believe, in the words of one website promoting the view, that "the hopeful alternative to the extinction of millions of species of plants and animals is the voluntary extinction of one species: *Homo sapiens*." They further argue that "when every human chooses to stop breeding, Earth's biosphere will be allowed to return to its former glory."

Theirs is an extreme form of a certain line of thinking, to be sure. However, weaker versions of it permeate our culture. There's a growing phenomenon of people deciding not to have children, or to have fewer children than they would otherwise, specifically due to fears of the additional degradation those children will inflict upon the planet. The reasoning is simple: All the recycling in the world and all the trash sorting you could ever do in all your life will never come anywhere *near* the impact of just not making as many humans. Not only do you prevent *their* lifetime of waste, but you prevent the waste of all their future progeny. Some argue that by having just two children, which is below the replacement rate, you're doing a good thing in terms of limiting future waste. But if everyone simply had two children, the population of the planet would gradually shrink to zero. Humans would eventually go extinct. If each couple just had one child, humans would be extinct in a mere six hundred years. Would the planet be better off without us?

As we saw in the chapter on carbon dioxide, scientist James Lovelock put forth a theory known as the Gaia hypothesis. It suggests that the entirety of all of the life on the planet functions like a single organism. Gaia, or the earth and its biosphere, is like a single living being that is formed of its composite parts, in much the same way as humans are single entities comprising billions of living cells. The cells can't perceive us any more than we can perceive Gaia. But we obviously exist, and perhaps Gaia does as well.

So, which is true? Are we part of the organism, or are we, by virtue of our sheer numbers and resource use, an invader, a virus? That's the question, isn't it? This "virus view" is at its core a belief in a moral equivalence of biological entities, that all life is of roughly equal worth, be it human or nonhuman animal. And to the extent that we humans, due to our excess population and consumption, are causing mass extinctions that have only just begun, then our existence, at least at our current population levels, is morally wrong.

Others reject this view. They maintain that a human life means something in a way that the life of a beetle doesn't, and thus we aren't a scourge but a blossoming. Before we came along perhaps there was paradise, but it was a meaningless paradise. Because no beetle ever marveled at the beauty of a sunset, this view goes, it's fair to say that the sunset wasn't actually beautiful until we came along to see it as such. But beyond that we've done marvelous things in this world, the likes of which no other creatures have even aspired to. We invented civilization, the rule of law, human rights, trial by jury, democracy, universal suffrage, and all the rest. William Shakespeare wrote *Romeo and Juliet,* Ludwig van Beethoven composed the Ninth Symphony, Michelangelo Buonarroti carved his *Pietà,* and J. K. Rowling created the world of Harry Potter. Point to a beaver that has done half as much.

Those who hold this view reject the characterization that they're burying their heads in the sand, pretending there aren't real, even potentially existential, challenges facing our planet. Instead they believe that human ingenuity is the key to working them all out. We may be the problem, but we are the solution as well.

So, are we a virus? That boils down to what you believe that people fundamentally are. Are we somehow unique and special, with each of our individual lives having inherent worth? Or are we just another species in a long chain of life, exactly like the millions that came before us, destined to thrive for a while and then face meaningless extinction?

The Worst Waster Ever

When it comes to waste, what's the most wasteful thing that has happened—ever? It's a difficult question to answer. In recent history, one might nominate the events of September 11, 2001, in which nineteen men with box cutters seized aircraft worth $250 million and used them to inflict $2.5 trillion of economic damage on the world. Further, the ripple effect of decades of war, loss of liberty, the time required of hundreds of millions of people to prevent such a thing from happening again, and the enormous diversion of resources necessary to prevent a recurrence make 9/11 a strong contender.

Going back further, one might nominate either of the world wars, which cost millions of lives and leveled entire countries. China's Cultural Revolution, Pol Pot's reign of terror, the Soviet gulags, and various exploitations by the haves of the world of the have-nots could all qualify. Further back in time, one might argue that chattel slavery would be a strong contender. Likewise, any of the practices we've already covered in the chapter on wasted human potential.

Let's narrow the question even further. What might be the most waste caused by a single person?

A good place to start might be the case of James Scott. In 1993, Scott and his wife were living on the Illinois side of the Mississippi River. His wife's job was on the Missouri side. One day Scott decided he wanted to party, and he didn't want the missus around. So he did something that seemed logical (to him, at least) and removed some of the sandbags that formed a levee next to the road she used to drive home, hoping to flood the street and strand her on the other side.

Regrettably, he overshot his mark and ended up flooding 20 square miles, destroying homes and businesses in the process. Scott's folly also washed out all the bridges within 200 miles. He was charged with "causing a catastrophe" (an actual legal offense), was found guilty, and was sentenced to life in prison. That was nearly thirty years ago. He's still in jail, and not eligible for parole until 2023.

When it comes to waste, true life is much stranger than fiction. The magnitude of Scott's waste is staggering. First, consider all the destroyed property. Second, contemplate the cost of incarcerating Scott for thirty years. And third, it's safe to surmise that Scott himself has things he would rather be doing than rotting away in prison.

As with many vignettes about waste, it's hard to find any winners at all. But James Scott merely merits an honorable mention for the title of Worst Waster Ever. Let's consider another candidate.

A hundred years before the birth of Christ there lived a Greek named Antipater of Sidon who compiled one of the earliest lists of the wonders of the world. He seems to have had no problems declaring his favorite. He wrote:

I have gazed on the walls of impregnable Babylon along which chariots may race, and on the Zeus by the banks of the Alpheus, I have seen the hanging gardens, and the Colossus of the Helios, the great man-made mountains of the lofty pyramids, and the gigantic tomb of Mausolus; but when I saw the sacred house of Artemis that towers to the clouds, those other marvels lost their brilliancy.

This temple that Antipater gushed over was actually the replacement for the original one, although the ancients maintained that it was a faithful reconstruction of the original. It was huge, not quite 100,000 square feet. It contained 127 columns, each 6 feet thick and six stories tall. Like all Greek temples, it had wooden rafters, and on July 21, 356 BCE, a man named Herostratus set it on fire, destroying it. Legend has it that the date of the destruction of the Temple of Artemis was the date of Alexander the Great's birth, and the Olympian goddess of the hunt was so busy with that blessed event she neglected to save her temple.

What caused Herostratus to commit this wanton act of violence? He confessed that it was simply so his name would be remembered for all time. Despite the best efforts of the Greek authorities, who in addition to killing him in a slow and painful manner decreed that his name never be mentioned or recorded, Herostratus seems to have succeeded.

What a waste! But as horrific as Herostratus's act was, he doesn't merit the title, either.

Let's consider Thomas Midgley, who lived from 1889 to 1944. While Scott and Herostratus each committed one horrific act, Midgley managed two acts so wasteful that they dwarf the imagination. The first, in 1922, was suggesting we add lead to gasoline, where it stayed in some cases until 1996 in the United States. By the 1970s, in this country alone, this practice led to the introduction of 500 tons a day of lead into our air, soil, and water. Researchers have credibly argued that this lead exposure, which is especially toxic to the brain, decreased the collective IQ of exposed children by multiple points, increased proclivity to violent crime, and caused a myriad of other problems including a variety of learning disabilities. From this point of view, the abolition of leaded gasoline in the United States is credited as *the* driving factor in the reduction of violent crime we have seen in recent years. (Leaded gasoline, by the way, is still used in many countries around the world.)

But Midgley didn't rest on his laurels, unfortunately. For his next act, he invented chlorinated fluorocarbons, or CFCs. CFCs seemed to be a miraculous substance, working wonders in refrigerators, inhalers, and aerosol spray cans by increasing their performance at very low monetary cost. He was universally lauded for this invention and did not live long enough to learn that CFCs were destroying the ozone layer, which protects the earth from harmful radiation. CFCs were largely banned by the Montreal Protocol in 1989.

However, even Midgley has competition for the title of Worst Waster Ever.

Perhaps the person who most deserves that distinction is none other than Genghis Khan, who famously said, "The greatest joy for a man is to defeat his enemies, to drive them before him, to take from them all they possess, to see those they love in tears, to ride their horses, and to hold their wives and daughters in his arms." While Conan the Barbarian uttered a paraphrase of this quote in the 1982 movie, it was Genghis Khan from whom he lifted it.

Khan began the Mongol conquests, which would ultimately kill as many as 40 million people—an astonishing 10 percent of the entire world's population. One might also give him "credit" for the additional 160 million killed a century later in the Black Death, which flowed from the Mongol lands west to Europe.

Khan killed so many people that vast tracts of Asia were depopulated and forest grew where people used to live, sequestering 700 million tons of carbon from the atmosphere. While that may seem like a good thing in today's climate, at the time those forests lowered the temperature of the earth, resulting in incalculable economic damage. And Khan's boast about holding in his arms the wives and daughters of those he slayed evidently wasn't poetic license: It's estimated that one person in two hundred alive today is one of his direct descendants. That's nearly 40 million people.

We should also award an honorable mention to the cat Tibbles, who lived a century ago. Tibbles lived with her owner, a lighthouse operator, on Stephens Island off the coast of New Zealand. Shortly after moving there, Tibbles began bringing her owner freshly killed wrens (which would later be named Stephens Island wrens, for obvious reasons). Unfortunately for the wrens, Tibbles seems to have single-handedly, or perhaps single-pawedly, brought about their extinction, as there were never many to begin with. Tibbles has some defenders, who claim that it isn't clear that Tibbles was the only cat on the island and that the wrens may not have immediately gone extinct after Tibbles's purge of them. But regardless, we still have to regard Tibbles as largely bringing about the extinction of an entire species all by herself.

Some Final Words on Waste

In the beginning of this book, we defined waste as a phenomenon that is subjectively bad and objectively avoidable. Then we refined that definition into something deeper and more philosophical, something inexorably tied up with the concept of values and purpose.

Is it possible to achieve a world *without* waste—one in which we exist at the ultimate pole of our continuum? After all, no amount of human effort can capture all of the energy of the sun. Nor can it abolish friction or bring back the dead. But the bits of the universe we do actually control can be put to constructive purpose, and to the extent we're able to align our values with our actions, we can live in a world with *less* waste.

How? As mentioned earlier, the complexity of the systems that make up our world makes it quite difficult to see how waste occurs; therefore we often don't know what to do to minimize it. There's cause and effect, certainly, but it's devilishly hard to tease out. Further complicating matters is the fact that many individuals and institutions in the world have a vested interest in doing things that are wasteful. It's big business for them, and they're often adept at hiring PR firms and lobbyists to further their individual agendas at the expense of the common good. In addition, reducing waste can be costly, and those on whom fall the costs of mitigating the waste are resistant to bearing them. This is a characteristic not just of large institutions but of individuals as well. We feel it whenever we're compelled to internalize our own externalities. Many of us are hesitant to give up the many conveniences that the modern world offers in the name of reduced waste.

All of these considerations are real barriers to reducing waste, but there's another, subtler one that lurks under them all. Often our strategy for dealing with waste is to change isolated behaviors on an ad hoc, piecemeal basis. We begin with "Plastic or paper?" when that should be our last question. The first question should be "What do we value?" since reducing waste inevitably comes with trade-offs. And often there is difference of opinion about whether the trade-off is worth it. What if, for instance, the least wasteful route to run a highway happens to be through a fragile wetland? What would you rather waste, the money to build a bridge over the wetland or the ecological diversity of the wetland? How do we as a society weigh the various trade-offs? Right now, we do it quite inefficiently.

We reduce waste when societies explicitly decide and then declare what they value, and then carefully build processes, systems, and traditions to actualize those values. The challenge is obvious: In the real world we have few absolutes to deal with. Would you reroute the highway if it spoiled an area of wetlands the size of a dinner table? Probably not. What if it was the size of a county? Probably. But where is the line? We lack a language that allows us to account for this nuance. For simplicity's sake we lump ourselves into groups that we broadly identify with, and we look with distrust at the motives of those who lump themselves into other groups.

Difficulties aside, what is the *mechanism* by which we can achieve a world with less waste? Who is it that "leads the charge," so to speak? There are six candidates we can examine: government, businesses, voluntary associations, institutional religion, public opinion, and individual action.

Government

The rationale for the modern bureaucratic state, with all its colossal and byzantine regulatory agencies, is not that people cannot govern themselves; rather, it's that individuals benefit by collectively ceding control over parts of our lives to highly specialized technocrats. We don't know if the drugs we're taking are safe, so we hire the FDA (or an equivalent agency in another country) with our tax dollars to make sure that they are. There's an alphabet soup of agencies to which we have collectively delegated control over parts of our lives in the interest of the common good as well as our own individual well-being. It's an imperfect and frequently frustrating arrangement, to be sure, but if we recall the old saying "Democracy is the worst form of government, except for all the others," we might come to a similar conclusion about the modern bureaucratic state.

In virtually all places that have high rates of recycling, it's achieved through extensive regulation. Low-flow toilets and the shift to LED lights came via government fiat, not individual action. In a world with inexpensive gasoline, carmakers made cars more fuel efficient not because they didn't have anything better to do with the billions of dollars that the R&D cost but because they were mandated to. We have cleaner air and water not because of a spiritual awakening among polluters but because polluters were forced into changing their ways by government regulators. The list goes on and on.

Libertarians object to these sorts of regulations, pointing out that if people want low-flow toilets or gas-sipping cars, they can buy them, so why inflict these product choices on the rest of us? Further, they point out that the higher costs that these requirements bring about disproportionately impact the poor, who are less able to bear this regulatory tax than the wealthy, including the privileged folks in Washington, D.C., writing the regulations. This cost can be substantial—one estimate is that 20 percent of the price of a new car factors in regulatory compliance. However, while libertarianism is certainly a philosophically consistent ideology, it's not one held by much of the population of the United States (or anywhere else, for that matter). A recent Gallup poll revealed that with regard to environmental regulation, 61 percent of Americans want the government to regulate more, even if it will slow economic growth. The proportion who think we need less regulation is 8 percent.

So, even though it's only in utopias that bureaucracies act solely in the interest of the public, it would be reasonable for government to take the lead in getting us to a world with less waste. The kinds of complexity that need to be sifted through on a thousand issues could be done so by each and every one of us, but why not delegate at least some of that tedium to government agencies? That certainly seems like a less wasteful approach.

Businesses

There's a delightful 1978 movie starring Warren Beatty called *Heaven Can Wait*. In it, football player Joe Pendleton is put into the body of an unscrupulous billionaire CEO named Leo Farnsworth, whose tuna company is netting and killing porpoises accidentally. At a board meeting, Pendleton/Farnsworth says they need to stop doing this, but a board member tells him that he isn't considering the expense. He replies, "But we don't care how much it costs, do we? We just care how much it makes. And if it costs too much, we charge a penny more. We make it part of the game plan: 'Would you pay a penny to save a fish who thinks?' "

Often when we see lists of ways we can each reduce waste, lower our carbon footprint, and decrease our consumption of electricity and water, the emphasis is on individual action. Perhaps that's not the place to focus. Maybe no person we know personally has ever made a water bottle or a plastic bag. We are collectively consumers of them, but the fact is that if Coca-Cola, which makes roughly three thousand plastic bottles a second, really decided it wanted to find a substitute, then a big impact would be made.

Similarly, when BP tweeted "The first step to reducing your emissions is to know where you stand. Find out your #carbon-footprint with our new calculator & share your pledge today!" journalist Andrew Henderson replied, "I pledge not to spill 4.9 million barrels of oil into the Gulf of Mexico." The comment is as true as it is biting. In point of fact, BP can do more to eliminate waste than Henderson could in a thousand lifetimes.

The challenge is that corporations are designed to make money. That's their raison d'être, and it has served society well. They provide goods and services that people want at prices they can afford. They invent the things that make our lives better.

Some warn that trying to weave a positive social agenda into corporations' DNA is misguided, as it's simply not their purpose in the world. However, we can't escape the fact that much waste happens in the economic sphere, and there corporations are in the driver's seat. To enlist corporations in the war against waste would require society to change its expectations of them. This is an ambitious goal. While there are a few notable examples to laud, earnings per share are all that really matters in most boardrooms.

Some argue that it would be better in the long run if corporations stay focused on maximizing profits, but within a framework of expanded government regulatory oversight that seeks to instantiate societal objectives within its statutes. This debate is far from being settled.

Voluntary Associations

Alexis de Tocqueville was a French diplomat and political scientist. He's known for a book he published in the 1830s called *Democracy in America*. It's a collection of his observations about the national character of the United States based on his travels around the young country. He wrote for a European audience, comparing the United States to the life they knew at that time. The book is still amazingly readable, and its observations about the United States still resonate. One of his observations was that in America, when people saw a problem, they turned for a solution not to the government but to what he called "voluntary associations"—what we might refer to as charities. He writes:

> Americans of all ages, all conditions, and all dispositions, constantly form associations. . . . The Americans make associations to give entertainments, to found establishments for education, to build inns, to construct churches, to diffuse books, to send missionaries to the antipodes; and in this manner they found hospitals, prisons, and schools. If it be proposed to advance some truth, or to foster some feeling by the encouragement of a great example, they form a society.

Doing so is still part of the American national character. In the United States, the vast majority of people donate to charities, and in aggregate donations amount to over $400 billion, or roughly 2 percent of GDP. That figure doesn't count the additional nearly $200 billion in volunteer labor that Americans freely give up.

In the United States today, there are about 1.5 million active charities. And you could see them helping us get to a world with less waste. The nice thing is that no one has to give people permission to start one. You see a problem; you start a charity. Whether you want to tackle waste by planting a million trees or picking up a million pieces of litter, you can start tomorrow. And many people in fact do. Consider Don Schoendorfer, who after being moved by seeing a person in a developing country unable to walk and without a wheelchair, invented a wheelchair that could be cheaply and durably made from materials he got from the local home improvement store. He started Free Wheelchair Mission and has given away more than a million of these chairs.

Undoubtedly, voluntary associations are part of the path to a world with less waste. Formed by passionate people who are close to the problem they're trying to solve, they can operate at a level of granularity and specificity that is hard for governments and corporations to match.

The problem with them is that they often lack the scale necessary to be really impactful on areas of the greatest waste. Further, all of their power is soft power; they can't compel the way governments can, nor can they redirect resources the way corporations can. Further, critics of them point out that they're inefficient in the extreme. Whereas governments simply tax and redistribute, charities risk becoming bloated, inefficient bureaucracies with little oversight, more concerned about prolonging their own existence than about serving the constituencies they were founded to help. In the worst cases, they become little more than tax shelters for the rich.

Institutional Religion

Religions would be natural leaders on the issue of waste, because of their distinctly ethical component. In the Pentateuch, which is regarded as holy by Judaism, Christianity, and Islam, humanity is given stewardship over the earth. This mandate can be interpreted as a broad endorsement of environmentalism and sustainable living.

Churches have taken up any number of social causes over the centuries, with varying degrees of success. In a world with waning church attendance and a rising number of people who don't identify with institutional religion, one might question how much influence the pulpits of the world might have on everyday actions relating to waste. With 80 percent of the world professing belief in a deity of some kind, it would be a mistake to underestimate the sway that organized religion could have in framing waste as a moral issue, not simply a financial one.

Public Opinion

Public opinion is one of the most powerful social forces on the planet. Consider our attitudes toward cigarette smoking. The seismic shift in public sentiment around that topic came about not because a generation that believed one thing died out and a new generation came up that thought differently. Rather, *the population changed its mind*. Within mere decades, smoking went from being the norm to being stigmatized. Readers in their forties and above might remember that people used to smoke everywhere—on planes, in restaurants, you name it. Then a few restaurants went all wild-eyed crazy and started offering "nonsmoking sections." A few years later it all flipped, and you had to request seating in the "smoking section." Eventually public opinion progressed to the point where officials could pass laws outlawing smoking without being summarily tarred and feathered. How did this happen? Public opinion changed.

How do you change public opinion? By shifting what's known as the Overton window. The Overton window is the range of politically acceptable policies currently in the mainstream. Smoking wasn't stigmatized all at once. The scientific proof of tobacco's danger wasn't sufficient enough. The conversation began in earnest with the idea that you shouldn't sell tobacco to kids. Why not have an age requirement? That seems reasonable. The Overton window shifted a bit. Shouldn't cigarette packs have warnings on them? What harm is that? Why not? The Overton window shifted a bit more. Can we ban smoking on short flights? Anyone should be able to go an hour without lighting up. The Overton window shifted again. And so on. Next thing you know, an indignant antismoking crusader is dumping a glass of ice water on you and your cigarette when you light up in a public park.

Of course, the Overton window shifts much more readily toward values that are held by a majority of the population. Can the Overton window be shifted on waste? To some degree it has already happened, in isolated areas. In the television show *Mad Men,* the Drapers embody the creators' vision of a typical American nuclear family circa 1962. A scene from the second season depicts the family enjoying a picnic in the park—and then casually shaking out their blanket, leaving a whole afternoon's worth of trash on the ground without a thought for what might happen to it, or of finding a trash can. To modern sensibilities, the episode seems almost sociopathic in its callousness.

Today most of us don't drop our refuse on the ground wherever we happen to be. Popular culture expresses disdain for conspicuous waste. The movie reboot of *21 Jump Street* featured Channing Tatum and Jonah Hill pulling up at their old high school in Tatum's 1970 Camaro—both the characters and the car clearly depicted as relics of the past. One of the cool modern kids sees the car and asks, "What's that thing get, ten miles to the gallon?" and Tatum proudly replies, "Nah, try like seven." The fact that poor fuel economy is a laugh line shows that the Overton window has moved significantly.

But the overall challenge with waste is that it's too nebulous.

Public smoking and littering are largely binary, but waste is a gradient where it's harder to draw clear lines. Nevertheless, shifting public opinion will be necessary to move into a world with less waste.

Individual Action

If you have any aerosol cans around the house, you may notice that on the rim of some of them is a red dot. That dot is designed to reduce waste. It lines up with the curved straw in the can and shows the corner the straw goes into. If you align the nozzle with that dot, you can get every last drop out of the can. Now that you know this factoid, you may well remember it and get just a bit more spray paint out of that can.

Thomas Jefferson is said to have advised that "the price of liberty is eternal vigilance." But that vigilance is also the price of a world with less waste. Can we expect everyone—or most people, for that matter—to remember a thousand more facts like the one about aerosol cans? Perhaps we can.

When all the complexity inherent in a world with less waste is listed out, it can feel a bit overwhelming. So, what can we do? Noam Chomsky has the following to say, and it's relevant here: "We have two choices: to abandon hope and help ensure that the worst will happen; or to make use of the opportunities that exist and perhaps contribute to a better world. It is not a very difficult choice." We can each individually do a great deal to get us closer to a world without waste.

If that statement seems unbelievable, take comfort in how far we've come already. Today's world has extraordinarily less waste than that of our forebears. Let's start with the small stuff. Do you remember life before GPS? How did you get from point A to point B? You probably used a printed map the size of a bedsheet that was outdated the day they printed it. Did it show the best routes? Did it tell you where there were accidents and slow-downs? If the map was going to tell you how to get to a place to meet a friend, how much time did you waste standing around before finding out later that your friend had a flat tire? How did you book a hotel before the internet? You picked a chain at random, called their 800 number, and asked if they had a hotel in Akron. Then you made a reservation. Was the hotel good? Or did the locals nickname it the "Stab and Stay"? Who knows? You just booked it, went there, and hoped that it wasn't Norman Bates who checked you in. Or how did you make an airline reservation? Do you remember printed tickets? They weren't that long ago. But let's move past the kinds of waste that we've made significant progress toward eliminating, which also includes war deaths, extreme poverty, and legalized slavery.

New technologies on the horizon can do even more to reduce waste. We will purify water more efficiently, generate clean energy more cheaply, and eliminate diseases more effectively. Every day scientists invent amazing new materials that promise to do miraculous things.

The good news is that many of these technologies disproportionately help those in the developing world. With better ways to grow food, provide clean water, and generate energy, the poorest billion or so people who have thus far seen little benefit from the technological marvels of the last couple of decades finally have a real reason to hope.

In the end, this is ultimately a book about hope. The world is full of waste, but there's less than there was yesterday, and tomorrow there will be even less. Sure, there's still too much, but we'll get there. We may not get to a world *without* waste, but it looks certain that we're going to get much closer.

Index

ABOUT THE AUTHORS

Byron Reese is a serial entrepreneur and author. He currently serves as CEO of JJ Kent. He is the author of *The Fourth Age* (2018), and *Infinite Progress* (2013), and he resides in Austin.

Scott Hoffman is CEO of International Literary Properties, which acquires literary copyrights and cash flows, and he owns and manages a portfolio of more than 1,000 titles. He is also one of the founding partners of Folio Literary Management, LLC. He has a BA from the College of William & Mary, and an MBA in finance from New York University's Leonard N. Stern School of Business. He lives just outside Austin, Texas.

ABOUT THE TYPE

This book was set in Minion, a 1990 Adobe Originals typeface by Robert Slimbach. Minion is inspired by classical, old-style typefaces of the late Renaissance, a period of elegant and beautiful type designs. Created primarily for text setting, Minion combines the aesthetic and functional qualities that make text type highly readable with the versatility of digital technology.

Also from Byron Reese
The Fourth Age

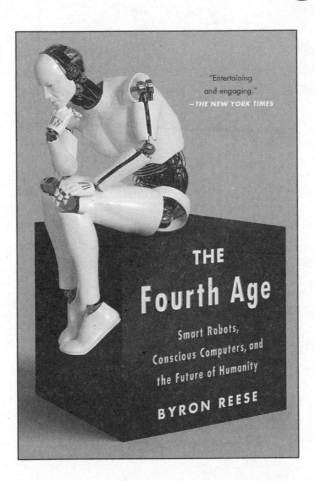

"Entertaining and engaging."
—*THE NEW YORK TIMES*

THE
Fourth Age
Smart Robots,
Conscious Computers, and
the Future of Humanity
BYRON REESE

Available wherever books are sold or at SimonandSchuster.com

ATRIA
BOOKS